HEALTH CARE DISPARITY IN THE UNITED STATES

AN URGENT CALL FOR UNIVERSAL HEALTH INSURANCE AND A PUBLIC HEALTH INSURANCE PLAN.

HEALTH CARE DISPARITY IN THE UNITED STATES
AN URGENT CALL FOR UNIVERSAL HEALTH INSURANCE AND A PUBLIC HEALTH INSURANCE PLAN.

Valiere Alcena M.D.M.A.C.P.

Books may be ordered through book sellers or by contacting Le Negre Publishing
(A subdivision of Alcena Medical communications Inc)
37 Davis Avenue
White Plains New York, 10695
Phone 914-682-8020
Fax 914-682-8066
www.dralcena.com
www.dralcena@aol.com

ISBN: (pbk) 978-0-9633365-4-5

Cover design by

Nick Nichols/Dr V. Alcena

Dedication:

This book is dedicated to the 46 million uninsured people in the U.S. and to all patients, physicians, and others who have suffered so unjustly and unfairly at the hands of HMOs since the inception of the HMOs program.

Contents

Preface

This book is written to bring to light the many wrongs that are going on in the health care insurance system in the U.S.

The U.S. health care insurance system is the most expensive health care system in the world and the most inefficient health care system of that of all the health care systems of other industrial countries in the world. The U.S. health care system is controlled by the HMOs. The HMOs run the health care system for financial gains as their only motives.

The CEOs of the HMOs device all sort of business models that enable them and their share holders to pocket the lion share of the 2.3 trillion dollars allocated yearly to pay for this inefficient and costly health care system.

They have zero interest in quality patient care.

The drug companies must also share some of the blames for the problems afflicting the U.S. health care system because, they make drugs that are so expensive and many people cannot afford to pay for them.

It is time for people in the U.S. to demand change and for the government to say enough is enough and come up with new legislation to create a brand new health care system to provide health insurance coverage for all American citizens.

Introduction

The U.S. health care insurance system is in a crisis and is broken. This system costs 2.3 trillion dollars per year and yet it is extremely inefficient. There are 46 million uninsured Americans and many million more with insurance who cannot receive timely and efficient treatments when they are sick. The HMOs stand between patients and their physicians to dictate when and what types of tests, medications, and treatments physicians can or cannot prescribe for patients. Physicians spend several hours per week for no pay, arguing with HMOs and their representatives to get needed care and services for patients. HMOs pick and choose whom to insure in their insurance companies.

They go to the greatest extent possible to prevent certain people from joining into their health plans to avoid spending too much money to provide health care for them. Evidence exists to show some minority physicians are not allow to join HMO panels using the excuse that these physicians are not Board certified. In fact the hidden reason is the fact that many of these physicians provide care for Blacks and Hispanics and other minorities, many of whom have serious chronic medical conditions that are very costly to pay for. By keeping these physicians off the HMOs panels, these patients are kept away from participating in the HMO plans.

In addition, for years many HMO companies used a fictitious Database –Ingenix owned by United Health to set payments for physicians and certain subgroup of patients. A law suit brought by the AMA, Medical Society of the State of New

York and others was settled recently with the help of New York State Attorney General to stop the HMO's involved from using this Database.

The HMOs have over the years devised many soft wares and business models to use to cheat and take billion of dollars from patients and physicians. The HMOs have been allowed to operate unchecked by both the Federal Governments and State Governments. The HMOs, because of their extensive wealth, have been in position to hire Lobbies to represent their interests in Congress and in state capitals all across the country. They have contributed many million of dollars over the years to politicians to influence them to support their way of doing business.

All these things are done while ignoring the health care needs of patients. Any one who dear to challenge the wrong doings the HMOs runs the risk of being black mailed by them.

For example, one of the largest HMO health insurance company doing business in New York , threw a physician off its panel of physicians by cancelling the physician's contract because, that physician stood up on behalf of a patient under his care when the patient was verbally abused by a physician on that insurance company's administrative staff.

That HMO insurance company took away more that 300 patients from that physician's practice and transfer them to physicians they did not know, who had no knowledge of their medical problems. Some of these patients had acute and life threatening medical problems. Many of whom spent a long time with no physicians to provide medical care for them. All these things were done to punish that physician for not being a "TEAM PLAYER". AND A BAD PERSON THAT MUST BE DONE AWAY WITH" The 300 plus patients they took away, had been the private patients of that physician for many years before that HMO recruited them to join their health insurance program because that physician's professional profile was outstanding. The patients joined precisely because they knew, trusted, and respected that physician.

It is the normal practice for and HMO contract to be given 90 days to be cancelled by either party to that contract. However, what that HMO insurance company which was the largest health insurance company in New York State did, was, it failed to reassign the patients that were taken away from that physician to other physicians, while still collecting their premium.

They got into major trouble, when one of these patients became seriously ill, and was taken to the emergency room of a local hospital; there were no private physician assigned by that health insurance company to take care of this gravely sick man.

The family called that physician and he went to the emergency room to provide the needed medical care for this patient to save his life. This incident took place over a Christmas holidays.

One of the many problems that powerful insurance company and its physicians executives ran into, was the fact, that 98% of the patients they took away were Blacks and Hispanics. It was very difficult for them to find physicians in the City of White Plains and surrounding communities who were willing to accept so many minority patients into their medical practices. As a matter of reality, these physician's medical offices are set up to provide medical services for white folks. Some of them may take a few people of color into their medical offices but by and large, these physicians practice medicine for whites.

Yes, the New York State Department of Health to its credit did take very serious punitive actions against this most largest and powerful HMO. Bravo! The little guy did win one this time, and, all the powerful DOCTORS who conspired to harm that physician who stood up for that patient subsequently lost their big fat jobs along with their big fat pay checks.

The one who actually signed the letter to remove the physician from the HMO's physicians panel was transferred from New York to far away Hawaii. It is not clear what happened to him after that. The one who signed the letter reporting the physician to the New York State Department of Health lost his job.

That HMO went further by reporting the physician in question to The New State health Department, in an attempt to have the physician's license taking away. That same physician was sent check by that HMO for a patient that he had never seen nor provided care for, all in an attempt to criminalize that physician. This cruel, vicious, and mean -spirited attempt failed flat on its face, but it cost money for the physician to hire a lawyer to defend him.

By their professional misconduct actions, they did to themselves that which they had hoped to do to that honest physician.

The HMOs are able to hire the best defense lawyers, and have been allowed to contribute money to judges running for offices.

Some of these HMOs behave like **GANSTERS** and use **GANSTER** like means to have their way. They have absolutely no respect for practicing physicians. They use practicing physicians as things to use to enable them to make money.

The HMO's call practicing physicians **PROVIDERS, to demean them.** If you are a physician and you are working for an HMO, you are called A PHYSICIAN.

If you are a practicing physician, the HMO calls you a PROVIDER. How regretful and how shameful!

It is now time to bring the HMOs to the carpet to account for all the wrongs they have done to destroy the U.S. health care insurance system because of corporate greeds.

The kidnapping of the U.S. health care insurance system could not have been possible without the help of many retired and present physicians, nurses, government agencies, and others who have benefited and are still benefiting financially from the HMOs.

Since the world began, money has had a tendency to corrupt. The cynical, uncaring, conscienceless, and heartless people who agree to work and sell their souls to the evilness of the HMOs are just as responsible for the mess that the U.S. health care insurance system is in, as are the CEOs and the share holders of the HMOs and other private health insurance companies.

CHAPTER 1

HEALTH CARE DISPARITY IN THE U.S.

A CRISIS BORN OUT MANY YEARS OF GOVERNMENTAL NEGLECT, INCOMPETENCE, AND CORPORATE FINANCIAL GREED.

AN URGENT CALL FOR UNIVERSAL HEALTH INSURANCE AND A PUBLIC HEALTH INSURANCE PLAN.

"Of all the forms of inequality, injustice in health care is the most shocking and inhumane." Martin Luther King, Jr.

"HEALTH CARE IS A RIGHT AND NOT A PRIVILEGE" Valiere Alcena, M.D.; M.A.C.P.

The U.S. health care system is in a crisis brought about by many years of governmental neglect, incompetence, and corporate financial greed of the Health Maintenance Organizations (HMOs).

The HMO industry was created in 1973 by congress during the Nixon administration. President Nixon authorized 26 million dollars toward the creation of 110 HMO projects. In 1973, the U.S. Senate passed a bill worth $5.2 billion to establish HMOs. The idea was to improve the nation's health-care delivery system by encouraging pre-paid comprehensive health-care programs.

The creation of HMOs was probably one of the biggest rouse and financial tricks that the U.S. population has had perpetuated on it in the 20th and the 21st centuries. What is called HMO is in reality "MANAGED COST". HMOs have two primary interests: First, to spend as little of the money they are collecting as possible and second, to fill their pockets with money and maximize profits.

There exists no evidence whatsoever that HMOs care about patients or the diseases from which they suffer. It appears that HMOs' principal capacity is to create computer software that shows them how best to take the most money possible from physicians, dentists, patients, hospitals, pharmacies, nursing homes, medical clinics and other health care related organizations.

There are about 2000 HMO and PPO health plans in the U.S. and altogether, they receive the lion share of the money that is spent every year in the U.S. on health care.

The health care industry is the second richest industry in the U.S. next to the military industrialized complex. The U.S. health care industry is 2.2 trillion dollars per year industry. Between 90% and93% of working individuals in the U.S. belongs to HMO's/PPOs. These insured workers pay a portion of the cost of their health coverage by having money deducted regularly from their pay checks with employers paying the rest. Employers typically choose the cheapest health plans they can find in order to save money. Employees therefore have no real chance to choose either good or the best health coverage plans. More often than not, the choice offered to them is the cheapest ones that provide the worst coverage. The CEOs of the employer companies are often working in concert with the CEOs of the HMOs/PPOs to arrange for the deal that is best for both entities with the interest of the employees being the most distant concern.

Many states in the U.S. have turned over their Medicaid health coverage to the HMOs in order to save money and to help balance their state budgets but at the expense of the poor men, women, children and disabled who live within their jurisdictions. All HMOs must be licensed by the individual state before they can do business in the state thereby creating a clear incentive for the HMO to do business on the state's terms.

The HMOs make their money - really their fortune - by establishing a business system that allows them to assemble a network of physicians, hospitals, medical clinics, pharmacies, etc. Enrolling these different entities enables HMOs to approach businesses of different types and sizes to submit bids for contracts for the enrollment of their employees into individual health plans. The HMO who offers the lowest bid usually gets the contract. In addition, HMOs have full, complete, and total control of which physicians are placed on their physicians' panel. HMOs utilize the different qualifications and specialties of the physicians, dentists, physical therapists, hospitals, etc. as a selling point to obtain contracts from corporate entities as well as individual states.

Many physicians who provide medical care for the poor in the inner cities and rural areas in the U.S. are frequently excluded from physicians (Providers) panel. HMOs have the well known secret that since poor people typically suffer from many serious and chronic medical problems that are costly to diagnose and treat, HMOs exclude many of these physicians as a way of avoiding the enrollment of these patients and their costly medical problems into their plans. To the HMOs, this is a business decision the entire purpose of which is to maximize profits by paying out as little money as possible while retaining the most money possible to share among themselves.

An example of the outrageous ways in which HMOs manipulate the system to cheat patients and physicians out of hundred of millions of dollars over many years, is the case of United Health. United Health, according to the Attorney General of New York State, created Ingenix Database and used information provided by that company, which was owned by United Health, to adjust payments which it made to patients and physicians for out – of –network medical payments. United Health, while not admitting any guilt, agreed to pay 350 million dollars to settle the case against it as part of an agreement with the Attorney General.

The law suit was brought against United Health by the American Medical Association, the Medical Society of the State of New York, and several other organizations. As part of the settlement agreement, United Health agreed to set up a new data base that will be controlled by an independent entity yet to be created. United Health contributed 50 million dollars to set up this fund, Aetna contributed 20 million dollars, CiGNA contributed 10 million dollars, Wellpoint contributed 10 million dollars, MVP contributed 535 000 dollars, HealthNow contributed 212,000 dollars and Independent contributed 475,000 dollars, for a total of 91 million dollars so far. Many other HMOs besides United Health used the Igenix Database to cheat physicians and patients.[1]

"U.S. Sen. Jay Rockefeller (D,W.Va.) has asked the Inspector General of the Office of Personnel Management to investigate whether federal employees were overcharged for health care because their plans use data base operated by the UnitedHealth Group subsidiary Ingenix. " I am concerned that federal employees participating in the Federal Employees Health Benefits Program may have been charged excessive out-of-pocket costs because of the health plans' use of the Ingenix data base products to determine rates for out of network services," Rockefeller wrote in a note dated March 31st. When testifying in front of the committee on Commerce, Science, and Transportation chairs by Senator Rockefeller, he told the CEOs of United Health and Ingenix: "I don't know, frankly, how you sleep at night." [2]

In 2006, the top eight HMOs in the U.S. earned $212,332,200,000. In 2007, these same HMOs earned $231,065,700,000, an increase in revenue of 5.4% to 27,7 %.[3] If the top eight HMOs earned almost 232 billion dollars, the rest of almost 2000 HMOs altogether must have earned more money than the top eight.

Using any criterion or set of criteria, the only way to describe this much money being made by persons who know so very little about good quality medical care is human indecency of the highest magnitude. The Federal government closes its eyes and allows this gross and outrageous system of HMOs to continue unabated. No doubt, the HMOs and the multi-billion dollars they earn every year for themselves and their shareholders, have highly paid lobbyists working in Washington to keep members of Congress working to maintain the status quo, as it relates to the HMOs' money making machine.

---

[1] Sources: 1.MSSNY, NEWS OF NEW YORK, Volume 64, Number 2, February 2009. 2. MSSNY, NEWS OF NEW YORK, Volume 64, Number 3, March 2009.

[2] Source: American Medical News, April 27, 2009.

[3] Source: AMA News February 25th 2008.

Some of the many problems that HMOs have created in their "managed cost" scheme include: a) paying physicians, particularly primary care physicians, between $30-$50 for routine examinations; b) paying between $65 to $70 for a complete physical examination; c) not permitting yearly physical examinations in most instances; d) paying between $4 and $6 for an EKG; e) paying only $2.00 for a blood sugar test; f) paying $1.50 for an urinalysis.

Some plans pay between $70 and $75 per day to treat a patient in the hospital. If the doctor sees a patient in the office for a serious medical problem and decides to admit the patient into the hospital, he or she will not get paid for the office visit, regardless of how much time he or she spends with the patient that day.

A recurring problem is that physicians are typically forced to go through a time consuming and demeaning process to obtain HMO approval of hospitalizations and outpatient tests medications for their patients. At any one time, an individual physician can spend between 45 minutes to one hour on the telephone speaking first with an operator, then to a nurse to discuss the patient's problem before any test or hospitalization can be approved. This is done sometimes for three or four patients during the course of a day, which may add up to three hours of unpaid work by the physician.

Frequently, after the long wait and exhaustive discussion, it is not unusual for the nurse to tell the physician that he or she cannot give approval to the request and she must refer it to the medical director. This process may take several days.

Clearly, the main job of the medical directors who work for the HMOs is to save money for the health plans by turning down physicians' requests for different tests and procedures.

All radiology examinations, stress tests procedures, and referral to specialists done without prior approval and referral by a PCP associated with a particular HMO will not be paid by the

HMOs/PPOs and the patients will be left standing with the bills, in some cases totaling thousands of dollars.

The same procedure applies to get certain medications approved for patients. HMOs contract with large pharmaceutical chains to provide medications for patients who are enrolled in their plans. They fight hard to get physicians to prescribe cheaper generic drugs so that they can save as much money as possible. The problem is that in many instances, both the quality and safety of these generic drugs are questionable.

While physicians and patients are caught up in this quagmire, the CEOs of these HMOs are earning million of dollars per year plus millions more dollars in stock options and other benefits.

The HMO managed cost system is set up in such a way that thousands of other health related entities are organized for the sole purpose of enriching themselves as part of the HMO grand business model. Those that are left to get the scraps off the table include:

Physicians and Physician assistants
Dentists and Orthodontists
Psychologists and Social Workers
Nurses
Physical therapists
Opticians and Optometrists
Podiatrists and Chiropractors
Laboratories and Laboratory technologists
Radiology facilities
Hospitals
Medical, Dental and Nursing Schools
Neighborhood Health centers
Nursing homes
Visiting Nurse and Home Care organizations
Home infusion organizations
Medical/surgical supply stores
Hospice organizations, etc, etc.

The Problem of the Uninsured

There are now 46 million uninsured individuals in the U.S. There were 87 million uninsured individuals in the U.S. in 2006 and 2008. [1] The U.S. is the richest country in the world and yet it is unable to provide quality health care to a large percentage of its citizens. Moreover, there is great disparity in the health care that is being delivered, mostly because of racism and poverty that is associated with it. Racial discrimination and religious bigotry are at the core of the many problems that afflict not just the U.S. but also other societies and are two of the major psychopathic illnesses plaguing humanity.

It is a great disservice and the height of hypocrisy that professional societies that are empowered to address issues of psychiatric illnesses have shown a reluctance to classify these behaviors for what they really are - mental illnesses. To fail to classify racism and bigotry against Blacks as a psychopathic behavior and a major psychiatric disorder is simply professional hypocrisy and dereliction of professional responsibility. It is a deeply inhumane and cruel thing for someone to discriminate against another human based on an immutable trait. The pain and suffering it causes cuts extremely deeply and leads to both mental and physical pain.

Racial discrimination is associated with:

1. Poverty
2. Poor education
3. Health illiteracy
4. High-unemployment
5. Welfare dependency
6. Drug addiction
7. Disparity in health care
8. Poor physical health
9. Mental illness and emotional suffering
10. Shortened life span, etc.

---

[1] (Source: Reuters March 4, 2009).

All of these factors add up to **"The Status Syndrome"**. The Haves and the rich have everything. The Haves not and the poor have nothing.

If you are rich in America and you get sick, you will have access to the best doctors, the best hospitals, the best machines to diagnose and treat your problems and the best and most expensive and effective medications. If you are poor, you are out of luck, period. That is what The Status Syndrome means, and that, quite simply, is how it works.

The U.S. health care system is top heavy with expensive medical technologies, unique medical machines, unusual and innovative surgical procedures. However, these things are set aside mostly for rich and powerful white people. Rich and powerful white men have been groomed from early childhood to believe that they are superior racially, economically, socially, and intellectually and therefore they inherently have certain entitlements. In their minds, their money can buy them the best health care possible and, quite often, they are right. Unfortunately and sadly, too many physicians and other allied health care professionals have given into these false and erroneous notions and time and time again have given special and preferential treatments to these nouveau rich. This way of doing things in America is not unique to the health care system: it mirrors what is happening in U.S. society as a whole. This top heavy method of delivering health care leads to a lot of wasteful spending and is partly responsible for the huge costs of health care in general- a problem that has been identified by President Obama in his recent proposals to reform the health care system.[1]

A good example of the waste of money that takes place in the U.S. health care system is the CT angiogram. On an average, the CT angiogram costs $1,500 per patient, and yet this test is no better than a coronary angiogram which costs much less money and in many ways, is a better test. Furthermore, a CT angiogram produces a lot of radiation which can lead to cancer. The CT angiogram machine costs about 1 million dollars to manufacture and operate. It has been observed that:" more than a 1,000 hospital and 100 private cardiologists either own or lease the CT

---

[1]   New York Times, June 14, 2009,

scanner". [1] The main reason that these physicians acquired this machine in the first place was to make a profit.

Another identifiable waste in health care spending in the U.S. is the cancer drug Avastin (bevacizumab) which costs $100,000 per patient per year, resulting in total sales of about $3 billion per year in the U.S. and $2 billion per year worldwide. Yet, according to the literature, this drug allows patients with advanced breast, colon and lung cancer to live only a few months longer than they would have lived without the drug.[2] No doubt when a patient is sick with a disease such as advanced cancer, the patient is fully justified in trying anything that offers some hope in a seemingly hopeless situation; that is human nature. But, must a drug be so exorbitantly expensive so as to put it out reach of most patients suffering from these dreaded cancers?

On the other hand, a good example of an expensive drug that works and may be worth the cost is Gleevec that treats chronic myelocytic leukemia (CML). Gleevec has cured many patients with CML, justifying the expensive cost of $377.20 for a 400 mg tablet- the usual starting dose per day for patients in chronic phase of CML. A 30-day supply of the 400 mg dose of Gleevec costs $3,772.00.

Still another example of the monetary waste in the U.S. health care system is the creation of CRITICAL CARE/INTENSIVE programs under the pretense that Intensive Care physicians are better able to provide care for patients admitted to ICUs/CCUs. According to an article in the June 3, 2008 issue, Vol. 148 of the Annals of Internal Medicine. The authors, who evaluated the medical records of 101,832 patients in 123 ICUs in 100 ICUs, found that hospital death rates were higher in patients who were cared for by so-called CRITICAL CARE/INTENSIVISE physicians when compared to those patients whose care was provided by regular Internists. This type of program has been in existence for the last decade and costs 55 billion dollars per year. What an outrageous waste of money! It is basically a money making scam that some medical directors and CEOs of some non- academic hospitals in the U.S. have used to favor one group of physicians against other groups who are not in league with them politically or socially. Internists are fully capable of providing competent and excellent care for patients

[1] New York Times, June 29, 2008.
[2] New York Times, June 6, 2008.

admitted to ICUs/CCUs in the U.S. Moreover, they are competent enough to know when it is necessary to ask for consultations from different specialists in the care of these patients.

The Role of the Pharmaceutical Industry

The Pharmaceutical industry in the U.S. is a 350 billion dollar plus per year (Brand drug) industry. The generic drug industry is more than a 25 billion dollar per year business and is increasing by the day. Most of the medications developed by the Brand drug industry have indispensable value in the treatment of diseases. Most medications are expensive to produce and, obviously, we cannot do without them. As such, physicians, as well as the public in general, appreciate the many research programs carried out by pharmaceutical companies. It has been said by some reports, that the drug companies spend somewhere around 19 billion dollars per year advertising their drugs to the public and physicians.

Drug companies advertise on TV, Radio, in Newspapers, and on the Internet. They also send drug representatives to doctors' offices trying to convince them to prescribe their medications. Drug representatives frequently arrange lunches and dinner meetings, all in an attempt to convince doctors to prescribe their particular drug.

The following is an itemization of the sale price of the top ten most prescribed brand-name drugs in the U.S. in 2007:

Lipitor: $ 8.1 billion
Nexium: $ 5.5 billion
Advair Diskus: $ 4.3 billion
Plavix: $3.9 billion
Seroquel: $3.5 billion
Singular: $3.5 billion
Enbrel: $3.4 billion
Prevacid: $3.4 billion
Aranesp: $3.2 billion

Epogen: $3.1 billion
Grand total: $41.9 billion[1]

---

[1] IMS Health Inc. (Internal Medicine News) Vol. 4, No 12, June 15, 2008

Pharmaceutical companies have gone global to increase their bottom line and bypass FDA regulations. As one example, ask the question: who is responsible for the safety of outsourced drugs? "Big Pharma outsources its products across the globe; approximately 80% of the active ingredients in U.S. prescription drugs are currently manufactured overseas."[1]

The FDA has stated that it will take thirteen years to inspect the plants where these brand medications are manufactured.

It is no doubt true that generic drugs are sold at a cheaper rate. However, while generic drugs are cheaper and more affordable, generic drugs are less supervised and thus much less controlled than brand-named drugs. The lesser supervision and control of generic drugs create justifiable doubt as to the quality of most generic drugs in the minds of both doctors and patients. Since September of 2008, the FDA has issued three separate alerts advising physicians as well as the public that 110 generic drugs on three separate lists are not safe to take. To make matters worse, some brand-name drug counterparts that previously appeared on these lists were no longer being manufactured, leaving the public to go without these brand-name medications that might have aided significantly in their treatment.

An example of such an occurrence is Toprol XL, commonly used to treat heart disease, the generic counterpart of which is Metoprolol ER. The manufacture has stopped making Toprol X, the brand name drug, while Metoprolol ER has recently been recalled by the manufacture.

HMOs like generic drugs because they cost much less money which is good for their bottom line. HMOs force doctors to prescribe generic drugs by constantly turning down their requests for brand drugs for their patients. It may take one to one and one half hours on the telephone to try to convince a pharmacy to grant brand drug pre-approval for a patient. I am of the firm belief that this process is made purposefully difficult to discourage doctors from making telephone calls on behalf of their patients. At the same time, physicians do not get paid for making telephone calls on behalf of patients, thereby reducing the physician's efficiency and, in the end, the physician's own bottom line. "The average

---

[1] Oncology News International, P. 35, Volume 17, Number 6/June 2008

physician spends 43 minutes per work day-more than three hours per week- dealing with health plan administrative requirements." Source: Health Affairs May14th 2009.The monetary loss that doctors suffer yearly must likely run into 31 billion dollars per year for all physicians in the country and $70,000 per individual physician per year, all for the benefit of the HMOs and their bank accounts. (Source: Health Affairs, May 14th 2009.)

In summary, HMOs enter into contracts with large pharmacy chains that are established specifically to provide the person different drugs the members of the plans need to treat their illnesses. These pharmacies are for-profit companies and, as such, in most instances, buy cheap uncontrolled generic drugs wherever they can find them which they then use to stockpile their inventory. This practice allows pharmacies to make the maximum profit possible. To effectuate this money making scheme, pharmacies, in collusion with HMOs, create a cumbersome system of pre-approval of medications that typically takes hours on the telephone by the physician just to get a particular medication approved for a patient. Such a system is clearly set up to discourage physicians from advocating on behalf of their patients to obtain clearly needed medications approved.

CHAPTER 2

THIRD WORLD HEALTH CARE IN A FIRST WORLD COUNTRY: THE MEDICAL CRISIS IN THE UNITED STATES OF AMERICA AND CONSEQUENTIAL HEALTH DISPARITIES

The United States is the richest and greatest country on earth and yet 46 million of its citizens have no health insurance. 16% of the U.S. population is uninsured. 20% of African Americans is uninsured and 33% of Hispanic is uninsured. In addition, 37.3 million U.S. citizens are living in extreme poverty. Fourteen million American children go to bed at night with nothing to eat and according to recent reports, at any given time; there are 3.5 million homeless people in the U.S. One out of every 50 children in the U.S. is homeless.[1]

In excess of 7.3 million people or 1 out of every 31 persons in the U.S. is in jail. There are 2.3 million men in jail in the U.S. more than 80% of whom are Blacks or Hispanics. Stated differently, one out of every 100 Americans is in jail, one out of every 15 Blacks is in jails, and one out of every 36 Hispanics in the U.S. being in jail. This leaves the Black and the Hispanic communities in an un-healthy and impoverished state with so many of their men incarcerated and not present to provide for their families, financially or otherwise.

---

[1]   U.S. Census Bureau Report, Tuesday, August 2008.

The unemployment situation in the U.S. is also in bad shape and is getting worse on a daily basis. The unemployment rate, however, among Blacks and Latinos is worse than any other subgroup, except Native Americans. The statistics are staggering: 13.4% of Blacks are unemployed; 10.9% of Latinos are unemployed and 7.8% Whites are unemployed as of March 12, 2009.[1]

Overall, Blacks represent 13.8% of the U.S. population and yet, 50% percent of the jail population is Black.

The U.S. health care system – a 2.2 trillion dollar industry in 2007 - is the richest in the world. The U.S. health industry is the second largest industry in the U.S. next to the military industrialized complex. "The U.S. spends $7.026 per person for health care per year, about twice the average in other industrialized countries without providing better care". [2]

The U.S. health care system costs twice as much as that of other health care systems in other industrialized countries. However, in comparison to other industrialized countries, health care in the U.S. is not as good. The U.S. mind set is to create new, big, and expensive machines as a way of demonstrating to the rest of the world its power in the health field. This superficial and artificial way of thinking runs throughout the fiber of American society and the American psyche. If it is American, it has to be bigger and better. Sadly, this is not always so.

The truth is, when it comes to health care, no machine can ever replace the sharp, wise, and experienced clinical mind of a medical clinician. In countries that lack expensive machines to help in the practice of medicine, the expert clinical mind rules supreme and provides better care for patients while keeping health care costs down. This proves that bigger and more expensive machines do not necessarily lead to providing better and more cost effective care for patients.

In 2006, The Commonwealth Fund published its first annual scorecard comparing the U.S. health care system to that of the top 10% health care system of other industrialized nations. The researchers used 37 indicators and found that the U.S. health care system scored only 66 on a 100 point scale. For example,

---

[1]   CNN, March 12, 2009

[2]   New York Times, June 29, 2008

the U.S. health care system scored 71 on Quality, 71 on Equality, 69 on long, healthy and productive lives, 67 on access, and 51 on efficiency in delivering health care.

According to a recent report, in 2006, "22,000 Americans died prematurely because they lacked health insurance. Lost productivity due to poor health and shorter life span could cost the economy as much as 200 billion dollars per year." More Blacks and Hispanics are uninsured than Whites. "The U.S. placed last among 19 industrialized countries in a study comparing preventable deaths". "The U.S. had the highest infant mortality rate and tied with New Zealand and the United Kingdom for the lowest average healthy life expectancy."[1]

Seventeen years ago, I authored a book that described the poor state of the health of Blacks in the U.S.A. and made recommendations for improvement.[2] The situation was dire then, but now the U.S. health care system is in a crisis and the crisis has a clearly disproportionate effect on Blacks and other minorities.

It is not surprising that Blacks, Hispanics, and Native Americans/Alaskan Natives bear the brunt of the disparities in the health care system in the U.S. because racism and cultural insensitivity are at the core of the health disparities that affect these groups. It is part of the American culture for the majority community to discriminate against minorities.

It has been commonly observed that when Blacks show up in the emergency room, they usually get less attention from the medical personnel. Blacks typically wait longer before they are seen in the emergency room. Their symptoms are paid less attention to and, in many instances; they receive poorer medical care than do Whites. It has also been observed that Blacks get admitted less often to the ICU/CCU. They are offered coronary angiograms less frequently and as a result, are offered coronary bypass surgery less often to treat coronary heart disease which may be life-threatening.

[1] Kaiser Family Foundation health care reform information: Www.health08.org and The Robert Wood Johnson Foundation health insurance reform information: http//covertheuninsured.org.

[2] Alcena, Valiere, M.D., F.A.C.P.: "The Status of Health of Blacks in the United States of America: A Prescription for Improvement", Kendall/ Hunt Publishing Company (1992)

If Blacks need an organ transplant, they are less likely to get it because the person making the decision is likely to be a 35 year old white resident physician. This is the reason that most organ transplants are given to white men, another glaring example of the disparity in access to even essential health care to preserve life itself.

Recently, an important report came out in the literature showing that when Blacks show up in the Emergency Room suffering with pain, they are less likely to be given prescriptions by White physicians for the pain; for fear that they might take the medications and in turn, sell them. The study confirmed that White patients in fact get prescriptions for pain by a much large number percentage than do Black patients.[1]

Now, in the prevailing economic crisis, whites and other ethnic groups are experiencing some of the same problems that minorities have experienced. So many people – of all races and ethnicities - have lost their jobs and are having problems paying their bills, feeding their families, etc. It has been reported that every 30 seconds, lack of health care causes someone to declare bankruptcy.[2]

Blacks and Latinos suffer more from the following diseases than do Whites:

1. Heart disease
2. Hypertension
3. Stroke
4. Cancer
5. High cholesterol
6. Obesity
7. Diabetes
8. Anemia
9. Asthma
10. COPD/Emphysema
11. Prostate cancer
12. HIV/AIDS
13. Glaucoma

[1]

[2]

14. Crack cocaine abuse
15. Depression

Poverty is associated with racism, poverty, poor education, health illiteracy, high un-employment, and welfare dependency.

The medical malpractice industry is also a multi-billion dollar industry that frequently engaged in bringing frivolous legal actions against doctors, which forces doctors to practice defensive medicine, thereby raising the cost of medical care.

IN THE MONEY

The seven largest health plans in the U.S. recently reported their earnings for 2007. Revenue was up for every company compared to 2006. California-based Health Net was the only one of the big insurers to post a drop in earnings from 2006 to 2007, due to paying three large settlements in class-action lawsuits. Revenue and income totals are in millions.

Company	2006 revenue	2007 revenue	Change	2006 net income	2007 net income	2006 earnings per share	2007 earnings per share	Change
Aetna	$25,145.7	$27,599.6	9.8%	$1,701.7	$1,831.0	$2.99	$3.47	16.1%
Cigna	$16,547.0	$17,623.0	6.5%	$1,014.0	$1,110.0	$3.43	$3.87	12.8%
Coventry	$7,733.8	$9,879.5	27.7%	$560.0	$626.1	$3.47	$3.98	14.7%
Health Net	$12,908.4	$14,108.3	9.3%	$329.3	$200.2	$2.78	$1.76	-36.7%
Humana	$21,416.5	$25,290.0	18.1%	$487.4	$833.7	$2.90	$4.91	69.3%
UnitedHealth Group	$71,542.0	$75,431.0	5.4%	$4,159.0	$4,654.0	$2.97	$3.42	15.2%
WellPoint	$57,038.8	$61,134.3	7.2%	$3,094.9	$3,345.4	$4.82	$5.56	15.4%

SOURCE: COMPANY FILINGS WITH THE SECURITIES AND EXCHANGE COMMISSION

Summary of the Revenue Earned by the 8 Largest HMOs in 2006 and 2007

- Total Revenue for 2006: $212,332,200,000
- Total Revenue for 2007: $231,065,700,000
- Change: Between 5.4% and 27.7% increase in Earnings

Source: AMA News February 25th 2008

All of these factors are taking needed money away from the health care system that could be better put to use to provide medical care for Blacks, Hispanics and other minorities.

People of color suffer from discrimination and health disparities in the U.S. Often times their symptoms are ignored frequently resulting in urgently needed diagnoses and treatments being delayed with resulting medical catastrophes.

The Disparity in the Availability of Minority Physicians

The solutions to correct these health disparities are many, complex, and difficult. However, one simple initiative that can be put in place relatively easily is to increase the number of Black and other minority physicians. This requires providing the necessary financial incentives for young black physicians to practice medicine where disadvantaged folks live and where the need is greatest.

There were 921,904 physicians in the U.S. in 2006; the breakdown among the various ethnic groups is as follows:

Whites: 514,254 =55.8%
Blacks: 32,452 =3.5%
Hispanics: 46,214 =5.0%
Asians: 113,585 =12%
American Native/Alaskan Native 1. 444 =0.2%
Others: 12,572 = 1.4%
Unknown: 201,383 =22%[1]
Source: Source: Physician Characteristics and Distribution in the
        U.S. 2008
Edition: American Medical Association.

In 2006, 1,335 Hispanics, 1, 219 Blacks, 3,425 Asians, and 11,190 Whites were accepted to medical schools in the U.S. It is evident that the number of Blacks and Hispanics accepted to medical school is small compared to the number of Whites and Asians.

The U.S. population is 306, 637,411

Blacks represent 13.8% of the U.S. population or 41, 1 million, Hispanics represent 15% or 46.9 million, and Asians represent 5.1% of the U.S. population or 15.5 million. Together, Blacks and Hispanics make up 28.4% of the U.S. population but only have 8.5% of the total physicians. Clearly, something is wrong and needs to be changed. It is crucial that the attitude and mind-set of physicians change and change rather drastically if

[1] American Medical Association News, Feb 04, 2008

there is going to be any real changes made to solve the problem of health disparities in America.

In addition, physicians in general need to be more racially sensitive toward minority patients and must be taught the concept of cultural diversity and what it really means so that they will be better able to understand the medical and social needs of patients under their care. It is imperative that medical schools include in their curriculum courses dealing with racial sensitivity and cultural diversity. Doing so will go along way toward solving health disparities.

The sum total of the racial, economic, educational, social, political and health disparities that are so pervasive and pernicious in the U.S. is that the median survival age in white men is 76 years, black men 70 years, white women 81 years and black women 77 years. The bottom line is that these disparities cause black men and women to die earlier than they would have died if the playing field in U.S. society were equal for all individuals.

The Importance of Universal Health Care

Finally, we need universal health care in the U.S. if we are really serious about eliminating health care disparities.

The root cause of racial discrimination against Blacks, Latinos and other ethnic minorities in the United States of America, is the disparity in the distribution of wealth that exists between Whites, Blacks, Latinos and these other minority groups. It has been reported that many years ago, the U.S. Government devised an economic plan to guarantee a permanent under class of Blacks, Latinos and other minorities in the U.S. Under this plan, "For every dollar of net worth owned by the median white family, the median African-American had only a dime and the Latino family had 12 cents." [1]

Owing to this plan, the Black and Latino population in the U.S. live in much greater levels of poverty than the White population. Fewer Blacks and Latinos have bank accounts. Fewer Blacks and Latinos own homes. Fewer Blacks and Latinos

---

[1] The Insight Center for Community Economic Development conference held in Washington D.C. on March 23 and 24, 2009 as reported on CNN during that same week.

graduate from high school. Fewer Blacks and Latinos attend college.

Fewer Blacks and Latinos have health insurance. More Blacks and Latinos go to the Emergency Room for medical care than do Whites. More Blacks and Latinos consume unhealthy foods than do Whites. More Blacks and Latinos suffer with chronic diseases than do Whites. Because of this consciously devised racist and unjust system, Blacks, Latinos and other ethnic minorities living in the U.S. are forever condemned to a permanent underclass with absolutely no hope of economic advancement which could free them bondage. The shackles of economic disparity is so tightly placed around the body and soul of Blacks, Latinos and other ethnic minorities living in the United States by Whites and the government, that the lives of these minorities are, in fact, shortened significantly.

The only way to reverse these disparities is to get rid of the old governmental policies that were designed to keep minorities in permanent social slavery/ and economic indenture and recreate a system that allows everyone to compete fairly by giving them an even playing field to work and compete. To achieve these goals will require a strong commitment on the part of government to allow minorities to get an education that will open the doors to get jobs that will increase opportunities to progress economically and socially. Education is the key to upward mobility in society.

The following chapters, which comprise the major portion of this book, will explore the manner in which disparities in the health care system are manifested in specific diseases in the U.S. population.

CHAPTER 3

DISPARITY IN HYPERTENSION IN THE U.S.

In 2009, there are an estimated 73 million individuals with hypertension in the U.S. That is, about one in every three persons in the U.S. has hypertension. Before age 45, more men than women suffer from hypertension. However, after ages 45-54, the percentage of men and women suffering from the disease is the same.

The incidence of hypertension in Blacks in the U.S. is higher than anywhere else in the world.

"From 1999-2002, the prevalence of hypertension in Blacks in the U.S. increased from 35.8 percent to 41.4 percent. During this same period, the percentage of hypertension went up to 44.0 in black women. The percentage of hypertension also went up in white men from 24.3 percent to 28.1 percent".[1] Blacks develop hypertension at an earlier age than do whites. Blacks have a 1.3 higher rate of non fatal stroke than do whites, a 1.8 higher rate of fatal stroke, a 1.5 times higher rate of heart disease related deaths, and a 4.2 times higher rate of end stage kidney disease than do whites.[2]

Based on an analysis by the Centers of Disease Control (CDC) of death certificates data from 1995-2002, Puerto Rican Americans had the highest hypertension-related death rate (154.0) and Cuban Americans had the lowest (82.5).[3]

---

[1] Archives of Internal Med. 2005, pp 2098-2104
[2] JNC 5 and 6 **[VAL: What is this reference??]**
[3] Morb. Mortal Wkly Rep. (MMWR) 2006: 55:177-180

In 2004, 54, 707 individuals died from hypertension in the U.S. Of this number, 23,099 were women and 31,608 were men. From 1979 to 2008, the death rates from the disease were 18.1 for white men, 51.0 for black men, 14.5 for white women, and 40.9 for black women. The direct and indirect cost of hypertension in 2008 was reported to be $69.4 billion.[1]

Hypertension is one of the most common diseases that afflict blacks not only in the U.S. but also in the world over. Hypertension is a disease that is associated with other diseases such as obesity, diabetes mellitus, and high lipid in the blood.

Worldwide about 1.5 billion people have hypertension, representing 1 in 4 individuals, causing 7 million deaths per year. In the U.S. presently 73 million individuals have hypertension. Hypertension is more common in blacks than whites; in fact 40 per cent of blacks in the U.S. have hypertension. Blacks constitute 13.3% of the U.S. population which translates into 41.1 million people, 16 million of whom have hypertension. Fifty percent of individuals 60 years of age and older in the U.S. have hypertension. About eighty percent of blacks who are 60 years old and older in the U.S. have hypertension.

In 2005, there were 73 million people in the U.S had hypertension. 34 million men and 39 million women had hypertension.

The prevalence of hypertension breaks down to 30.6% of white men, 41.8% of black men, 44% of black women, and 27.8% of Mexican American men. The prevalence of hypertension in Hispanics or Latinos is 18.2%. The prevalence of hypertension in Asians is 16.7% and the prevalence of hypertension in American Indians or Alaskan Natives is 21.2%. Pre-hypertension is a condition that exists when the systolic blood pressure is 120-139 and the diastolic pressure is 80-89. There are 69.7 million people in the U.S., 37.4 million men and 27,800.000 women with pre-hypertension.[2]

The total number of deaths attributed to hypertension in the year 2002 was 261,000. The total yearly mortality due to hypertension in the U.S. is 49,700. Of the yearly mortality due to

---

[1]   NCHS Compressed Mortality File: underlying causes of death, 1979-2008.-

[2]   Med. Sci. Monitor 2005:11: CR403-409.

hypertension in the U.S., 20,540 occur in men, 14,700 occur in white males, and 5,300 occur in black males.

In 2002, the death rate from hypertension in the U.S. was 17.1 per 100,000. The actual death rates for white males who died because of hypertension were 14.4 per 100,000 and the death rates for black males who died because of hypertension were 49.6 per 100,000. Of individuals who have hypertension with elevated blood pressure, 68% of them are receiving treatment and only 64 % of those who are receiving medications for hypertension have their blood pressure well controlled.[1]

Metabolic Syndrome is the combination of the following diseases: hypertension, obesity, diabetes and hyperlipidemia. The Syndrome is frequently associated with essential hypertension in both men and women. In the U.S., an estimated 47 million individuals have Metabolic Syndrome. Metabolic syndrome in men and women is associated with the following conditions: a waist circumference greater than 40 inches, fasting blood sugar of 110mg/dl or higher, high density lipoprotein (HDL) cholesterol less than 40mg/dl, blood pressure of 130/85 mm Hg or higher, and serum triglycerides level of 150mg/dl or higher.

Mexican Americans have the highest prevalence of Metabolic Syndrome - 31.9%. Whites have a prevalence of Metabolic Syndrome of 23.85%, blacks have a Metabolic Syndrome rate of 21.6%, and those who refer to themselves as others in the U.S. have a prevalence of Metabolic Syndrome of 20.3%. Complications associated with these diseases commonly cause severe suffering and even death.

What is hypertension?

Hypertension occurs when the systolic part of the blood pressure is higher than normal and the diastolic part of the blood pressure is higher than normal.

What is the systolic part of blood pressure?

The systolic part of the blood pressure is the upper part of the number in the blood pressure reading machine.

---

[1] CDC, Data Brief 2005-2006.

What is a normal systolic blood pressure?

A normal systolic blood pressure ranges from 100 to an upper limit of 139.

What is normal diastolic blood pressure?

A normal diastolic blood pressure ranges from 60 to an upper limit of 89.

CLASSIFICATION OF BLOOD PRESSURE IN ADULTS AGE 18 YEARS AND OLDER

Category	Systolic mm/Hg	Diastolic mm/Hg
Normal	100 – 130	60 - 85
High Normal	130 - 139	85 - 89
Hypertension		
Stage I	140 – 159	90 - 99
Stage II	160 – 170	100 -109
Stage III	180 - 209	110 -119
Stage IV	210	120 or greater

(Archives of Internal Medicine, Volume 153, January 1993)

New blood pressure Classifications

Classification	Systolic		Diastolic
Normal	<120	and	<80
Pre-hypertension	120-139	or	80-89
Stage 1	140-159	or	90-99
Stage 2	160+		100+

Source JAMA Volume 289, No 19 May 21, 2003 The JNC-7 Report

A multi-institutional study which was published in the Proceedings of the National Academy of Sciences in 2002 and which took 18 years to be conducted, showed that the kidneys of American Caucasians, Japanese, and Ghanaians contain G protein coupled receptor kinase type 4 in the same amount in all three racial groups. The study confirmed that this protein causes salt retention by the kidney in all three groups. Hence, the basis of essential hypertension was discovered.[1] The study confirmed that salt retention in the human body and salt sensitivity is the genesis of essential hypertension regardless of ethnicity.

This lifesaving gene was a necessity in the old world in Africa where human life began, but is a detriment to health in the New World. The salt-sensitive gene is extremely strong and highly penetrating. Diet contributes to several of the most common diseases. The interplay of hypertension, diabetes mellitus, obesity and high cholesterol, referred to as metabolic hypertension, or Syndrome X, is quite common. All four components of the disease are genetically transmitted. Presently 48.6% of African men are obese in this country. When a black baby starts out in life with that abnormal genetic package, by the time that infant boy grows into full adulthood and has to go through all the psychosocial, social and economic stresses to survive in the present society, he is certain to suffer from the adverse effects of metabolic hypertension.

The history of salt sensitivity and secondary fluid retention resulting in elevation of blood pressure did not start thousands of years ago as a disease but rather as a God-given measure, to maintain life and prevent death, as described below.

Living conditions in ancient Africa thousands of years ago where the human race began, and to a significant extent in present-day Africa, were quite harsh with men working in extremely high temperatures. Under these conditions, the human body loses a lot of salt through the skin and, in so doing, loses water along with salt through the skin as sweat. Wherever salt goes in the human body, water goes with it. When a person loses salt and too much water with it, the body can become dehydrated quickly. Once the intravascular system is depleted of fluid the body risks

---

[1] **Proceedings of the National Academy of Sciences (2002; 99: 3872-3877)**

being collapsed. It takes between 1800 cc to 2500 cc of fluid lost ordinarily to cause the blood pressure to fall in a 70 kg person or a normal-sized person. Once the kidneys sense that the blood pressure is falling, their normal tendency is to prevent salt from going out of the body in the urine, thereby attempting to maintain the blood pressure in the normal range. Through that mechanism, salt remains in the body and keeps water with it to maintain blood pressure and to prevent the body from collapsing. The kidneys are able to do this because there are special genes that are located in the kidneys that enable them to hold on to salt.

All the billions of people who live across the world have to face the fact that they suffer the same fate of salt sensitivity when it comes to the problem of hypertension as do those who are clearly African in appearance. Those who choose not to acknowledge any connection with their African ancestry most commonly do so for psychosocial, social, and economic reasons while disregarding their health related commonalities, often to their detriment.

However, when it comes to the scientific facts, they are clear and obvious, because the African gene is quite penetrating, regardless of the percentage, degree and extent to which one is associated with it from a genetic (DNA) standpoint. The gene will clearly show up when it comes to certain diseases that are of a hereditary nature.

People in the New World have largely had the syndrome of salt sensitivity to thank for their existence. That is so because when the original human beings left Africa, they traveled on wooden wharfs on the open sea to come to what is now the new world and one can imagine the adversities they had to overcome, including dealing with the hot sun and storms of different intensities, as well as hunger and thirst. Those who survived did so in part because their kidneys were able to hold on to enough salt, which kept some water in their bodies and so prevented death from dehydration. The kidneys of these men and women contained a gene (known today as G protein coupled receptor kinase-4) which enabled them to hold enough salt in their bodies to hold on to water and which kept them alive. We must thank this gene to a great degree for our existence.

The immediate forebears of blacks in the new world would never have survived the voyage, the horrible voyage of torture and brutality across the Oceans, to arrive in the Americas and elsewhere. The fact that their kidneys were able to hold on to enough salt and water kept them alive. Blacks and other races, by whatever names they may wish to call themselves and who carry the salt sensitivity gene in their kidneys (and all humans do so to one degree or an other), owe our African ancestors a debt of gratitude for our existence.

However, along with all the good qualities that the forebears of the human race have passed on, they also passed on the gene for hypertension. Hypertension and its associated complications have caused and are presently causing more deaths than any other disease in the United States as well as worldwide.

Blacks and Asians have hereditary low renin or high volume hypertension. Blacks have low renin because the high salt content in their bodies suppresses renin production. Asians have low renin because their high salt diet suppresses renin production thereby causing the level of renin to become low. This is very important clinically because they are certain medications such as beta- blockers and simple angiotensin converting enzyme inhibitors that work to lower blood pressure only when the renin level is high. Therefore, prescribing these types of medications to individuals in these two racial groups- Blacks and Asian - to lower blood pressure may not be effective.

Hypertension affects the heart, brain, kidneys, and eyes, the so-called end organs. Hypertension can cause atherosclerotic plaques to develop within their coronary arteries, resulting frequently in heart attacks and death. Hypertension also causes the heart to become enlarged because the heart has to pump against a high load - high blood pressure and, over time, the muscles around the heart become hypertrophied, resulting in enlarged ventricles.

Once hypertrophy sets in, because the heart muscle only has a finite length to which it can be stretched, it reaches the point that it can no longer stretch. Then it begins to pump ineffectively and the ineffectiveness of the heart muscle resultss in what is referred to as cardiomyopathy with secondary congestive heart failure. At that point, the heart is unable to push the blood away from the ventricles (heart chambers) and the blood/ water backs up into the lungs and accumulates as fluid. Then, shortness of breath develops. If not treated quickly it can result in what is referred to as pulmonary edema (acute congestive heart failure), the result of which, when it is not treated quickly and acutely, is immediate death. In the less dramatic way, the enlarged heart sets in and the affected person begins to develop lassitude, inability to walk down the block without stopping several times, inability to sleep at night on one pillow and constant coughing at night. That condition is referred to as nocturnal coughing. All these are signs that the heart is failing. If the person gets to a physician quickly, the condition can be diagnosed and treatment can be started with appropriate medication to prevent this acute condition from occurring.

Another organ that suffers frequently from the effect of hypertension is the kidney. Hypertension damages the kidney resulting in kidney failure. The way this happens is that the pressure rises within the vessels which run through the substance of the kidneys. All the different tissues of the kidney need blood vessels of different sizes to carry blood and oxygen to them. As the pressure rises within the kidneys, there are structures within the kidneys referred to as glomeruli (the filtering system of the kidneys) which need to be fed blood and oxygen. As the blood pressure rises, these very delicate blood vessels begin to rupture. They are rupturing without the affected person realizing what is occurring. After a while, these vessels rupture and die out and the tissues to which they are responsible for bringing blood and oxygen will no longer be there. Consequently, these areas of the kidneys die. Eventually, the affected person loses so many glomeruli that the kidneys cannot function properly (known as renal insufficiency). Once the glomeruli die, the kidneys can fail suddenly. Once the kidneys fail, waste materials accumulate within the body, resulting in swelling of the legs with smelly breath and salty skin. A condition referred to as chronic renal failure with uremia also develops. At that point, either peritoneal dialysis or hemodialysis on a chronic basis must be used to clean the blood free of toxic materials to maintain life. If a person is fortunate enough to get a kidney transplant, and the transplant succeeds, then he can go back to normal kidney function and a normal life.

In summary, high blood pressure that goes untreated can damage the kidney to the point of kidney failure. Typically, the kidneys fail slowly, losing function in a gradual way, so that a person who is not having regular medical check-ups may not have the disease detected until it is too late.

Another organ that is very sensitive to the effects of hypertension is the eye. When the blood pressure rises in the body, the pressure also rises within the vessels in the eyes. The vessels inside the eyes are quite fragile and so they can be damaged quite easily. The damage that occurs in the vessels inside the eyes causes different degrees of leakage to occur. If left untreated, blindness usually results. Hypertension is also associated with an increased incidence of glaucoma, a common disease in particular in people of color.

The brain is another organ that suffers the effects of hypertension to different degrees. Over time, the effects of elevated pressure cause plaques to develop within the small and large vessels of the brain. The damage that occurs within the small vessels results in multiple small vessel infarctions inevitably leading to the condition referred to as "multi-infarct syndrome". Multi-infarct syndrome is the most common cause of senility in men also knows as organic brain syndrome. This condition is common due to the high incidence of hypertension which frequently goes either completely untreated or improperly treated. Of the roughly 73 million individuals who suffer with hypertension in the United States, only about 40 percent seek medical attention. Sadly only about 20 percent of those seeking medical attention get their blood pressure controlled to about 140/90, and 140/90 is not considered good blood pressure control.

People of color are generally more prone to the development of early senility due to untreated hypertension or poorly treated hypertension. Elevated blood pressure can cause three different types of major strokes to occur (also knows as a cerebrovascular accident): ischemic stroke, hemorrhagic stroke, and embolic stroke.

Ischemic stroke occurs because of the chronic narrowing of the affected vessel with plaques and/or the rupture of plaques within the affected vessels, resulting in bleeding. The ensuing clot formation acutely closes off the vessel, cutting off blood flow, resulting in what is known as a stroke.

Elevated blood pressure causes hemorrhagic stroke to occur due to chronic damage to the affected vessels, resulting in acute rupture of those vessels and causing hemorrhage to occur inside the brain.

Hypertension-associated embolic stroke can occur because of hypertensive heart disease with enlargement of the heart. This can cause atrial fibrillation to develop, and if the atrial fibrillation remains untreated with anticoagulants such as Heparin or Coumadin to prevent clot formation, then the clot might become dislodged from the atrium to the brain, causing an embolic stroke to occur.

Regular exercise, abstinence from smoking and the abuse of alcohol will decrease the incidence of high blood pressure. These preventive measures can lead to decreasing some of the adverse consequences of high blood pressure such as stroke, heart attack, and kidney failure. All of these factors contribute to black males' decreased median survival of 70 years as compared to the white males' median survival of 76 years and black females' decreased median survival of 77 years as compared to white females' median survival of 81 years.

Obviously, a change in lifestyle is necessary to help decrease the incidence of hypertension. However, a change in lifestyle is not always realistic in the lives of people of color considering the often poor economic circumstances of Blacks, Hispanics, and Native Americans in the United States. The stress brought on by a multitude of problems associated with poverty and racism plays a major role in the elevation of blood pressure in minorities. This is especially unfortunate given that Hypertension is one of the leading causes of morbidity and mortality.

Essential hypertension is the same disease in all people without regard to race or creed. This is because everyone who suffers from essential hypertension is salt sensitive to one degree or another and must be given a thiazide diuretic as the main medication. Doing otherwise guarantees the development of major difficulty in controlling their blood pressure. Treating high blood pressure with a proper regimen of medication goes a long way in decreasing death rates by preventing the multitude of hypertension-associated complications. Treating blacks, Asian, and people of color in general who are likely to be salt sensitive and have low rennin is essential with the proper second anti- hypertensive or, when necessary, a third anti-hypertensive medication such as Calcium channel blocker or Angiotensin-II Converting Enzyme Antagonist.

If it is reasonably expected to properly control hypertension as it commonly occurs in the general population, the medical community must stop fostering the covert false notion of **racialism** in its treatment. Hypertension is colorblind. It knows no skin color or racial differences and affects every one in the same adverse way owing to the disease's genesis.

Hypertension is an easily treatable disease if the right medication or medications are provided for the right people under the right clinical circumstances.

CHAPTER 4

Disparity in stroke in the U.S.

Stroke is one of the leading causes of death in the U.S. Every year, about 780,000 individuals have a stroke in the U.S., 600,000 of which are first strokes and 180,000 of which are recurrent strokes. About every 40 seconds somebody has a stroke in the U.S.

Every year, 60,000 more women have strokes than men do.

The risk of first strokes is twice in blacks as compared to whites.

Eighty eight percent of these strokes are ischemic, nine percent are intracerebral hemorrhage, and three percent are sub-arachnoids hemorrhage.[1]

One of every 16 deaths in the U.S. is due to a stroke. In 2004, 150,074 persons died of strokes, 58,800 of them were males, and 91,274 of them were females.

Blacks have a higher prevalence of small vessel strokes than do whites.

The most common risks for stroke are:

1. Hypertension
2. Diabetes mellitus
3. Obesity
4. Arteriosclerosis
5. Hyperlipidemia (high cholesterol)
6. Tobacco smoking

[1]   Heart Disease and Stroke Statistics -2008 Update, American Heart Association.

7. Atrial fibrillation
8. Primary polycythemia
9. Secondary polycythemia
10. Essential thrombocythemia
11. Sickle cell anemia
12. Hypercoagulable state
13. Thrombophilia
14. The Obstructive Sleep Apnea Syndrome
15. Serum B12 deficiency
16. Serum Folic acid deficiency
17. Elevated Homocysteine level
18. Decreased protein C level
19. Decreased protein S level
20. Decreased anti thrombin III level
21. Taking birth control pills
22. Taking any estrogenic hormone
23. Elevated Lipo-protein-a level
24. Elevated anti-phospholipin antibody
25. Elevated circulating lupus anticoagulant
26. Hematocrit level 40% or greater in patients with chronic renal failure can cause stroke to occur
27. Hematocrit level greater than 34% in patients with sickle cell anemia can cause stroke to occur
28. Hyperviscosity state in patients with multiple myeloma can cause stroke to occur
29. Nephrotic syndrome
30. AIDS
31. Factor V Leiden mutation
32. Prothrombin G20210A mutation
33. Migraine with aura

The percentage of the most common types of strokes occurring in the U.S. is:

1. Atherosclerotic or ischemic stroke—88%.
2. Embolic stroke, which represents 24%.
3. Hemorrhagic stroke represents ----10%.

4.   Subarachnoid hemorrhage—3%.

Other frequently occurring types of stroke include: ruptured aneurysms, bleeding arteriovenous malformations, transient ischemic attacks, occlusion of carotid arteries by plaque causing stroke to occur because of lack of blood flow to the brain, and Lacunae stroke.

Atherosclerotic disease of the brain occurs in a very large percentage of individuals. The reasons for this high percentage of brain atherosclerotic disease include:

1.   Hypertension
2.   Obesity
3.   Diabetes mellitus
4.   Hyperlipidemia/high cholesterol
5.   Tobacco smoking
6.   Alcohol abuse
7.   The different stresses associated with life including discrimination of different types, poor economic conditions, poverty, lack of education, poor health, etc., and all the other bad things that most people of color have to cope with, which include poor quality foods rich in fats, salt and carbohydrates. All these negative factors can contribute to the development of a stroke.

The human brain is in total control of all activities associated with being human. The ability of the brain to control what humans do differentiates the human animal from all other animals.

The following is a partial list of some of the skills that the human brain controls:

1. The ability to think
2. The ability to gather information and process such information logically and rationally to formulate judgments rightly or wrongly.
3. The ability to breathe
4. The ability to see
5. The ability to hear

6. The ability to smell
7. The ability to feel
8. The ability to taste
9. The heartbeat and other crucial functions of the heart
10. Lung functions
11. Hunger
12. Lack of desire for foods
13. Thirst
14. Lack of desire to drink
15. Sleep
16. Insomnia
17. Happiness or unhappiness
18. Moods such as Elation or Despair
19. Motivation of Lack of Motivation
20. Hardworking habits
21. Laziness
22. Neatness
23. Sloppiness
24. Anger
30. Aggressive behavior
31. Antisocial behavior
32. Pleasant and friendly behavior
33. Lying as a habitual behavior
34. Honesty
35. Dishonesty
36. Criminal behavior and other antisocial behaviors
37. Sexual orientations/preferences
38. Sexual desires
39. Erectile function
40. Ejaculatory functions and sexual satisfactions
41. Bowel functions
42. Urinary functions
43. Chewing
44. Swallowing
45. Sneezing
46. Coughing

47. Yawning
48. Lying down
49. Sitting
50. Bending
51. Standing
52. Walking
53. Running
54. All other motor body functions
55. Writing
56. Reading
57. Speaking

In short the brain is in control of all bodily functions

Different parts of the human brain are in control of these different functions, so when the brain is damaged by a stroke, accidents of one type or another, by infections or other abnormalities that interfere with its normal functions, one or several of these vital functions become impaired in one way or another.

Hypertension causes strokes through three basic mechanisms: 1) Increased blood pressure in the vessels within the brain causing the inside part of these vessels to become damaged, and over time, the damaged areas of these vessels trapping platelets and other material as they pass through with the blood. A nidus of these different materials develops within these vessels and the result is plaque formation. The formation of plaques within these vessels leads to narrowing of these vessels, impeding blood flow. Superimposed on the plaque frequently is a clot which can acutely close off a vessel, resulting in a cerebrovascular accident stroke. A plaque within a vessel can cause a stroke through different mechanisms:

(a) The plaque can cause the vessel to become narrowed impeding blood flow and oxygen delivery to a particular part of the brain.

(b) The plaque that sits inside that vessel can break off, causing either an embolus or a clot to start forming, resulting in a stroke as has just been outlined.

2. Another mechanism through which hypertension causes stroke is acute intracerebral bleeding secondary to very elevated blood pressure causing rupture of a blood vessel, resulting in bleeding within the brain. Bleeding inside the brain can result in a coma because of edema (swelling) within the brain, and if the coma lasts too long, then the result can be death of the affected person.

3. Another type of stroke syndrome that occurs in individuals who have been hypertensive for a long time, and in particular if the blood pressure has not been treated or not treated properly as aforementioned, is multiple small vessel infarctions of the brain.

These small vessels are located deep inside the brain and supply blood to very vital structures within the brain. This condition is associated with early memory loss resulting in organic brain syndrome. Multiple small vessel infarctions are second only to Alzheimer's disease as a cause of senility. In fact, it is probably more common than Alzheimer's in terms of causing senility because there are so many more people who are hypertensive than those who have Alzheimer disease. It is not uncommon to see a 40-year-old person of color who has been hypertensive since his 20s or 30s having difficulty remembering very simple things because of multiple small vessel infarctions of his or her brain, as disclosed on brain MRI (magnetic resonance imaging). (CT of the brain does not show that syndrome very well.)

Stroke is a very disabling disease that kills more than 150,000 people every year in the U.S. In addition, stroke causes major disabilities in those who are affected resulting in incredible sufferings and loss of income. In 2008, the total cost of stroke in the U.S. was $65.5 billion.

CHAPTER 5

DISPARITY IN HIGH CHOLESTEROL

There are 106.7 million people in the U.S. with high cholesterol. There are 50.8 million men and 55.9 million women with high cholesterol. The cholesterol is high if it is 200 milligrams per deciliter or higher. In the U.S., 37.2 million individuals have total blood cholesterol 240mg/dL or greater. 17.2 million men, and 19.9 million women in the United States have blood cholesterol at that high level.[1]

In 2004, 869,724 individuals died of cardiovascular disease (CVD) in the United States making CVD the number one cause of death in the U.S. In the U.S., 36.3 per cent of the 2,397,615 deaths that occurred in 2004 were due to CVD. Worldwide, 17 million people die of CVD every year. High cholesterol is one of the risk factors for coronary artery disease (plaques inside the vessels around the heart), and strokes, two of the triad that makes up CVD. Other risk factors include:

1. Hypertension
2. Hyperlipidemia
3. High triglycerides in the blood
4. Elevated low-density lipoprotein in the blood (LDL)
5. Low high-density cholesterol in the blood (HDL)
6. High cholesterol /LDL ratio in the blood
7. Family history of early heart attack, especially where a parent died of heart attack in his or her early 40s to mid-50s
8. Cigarette smoking

[1] Source: NCHS and NHLBI

9. Diabetes mellitus
10. Elevated lipoprotein A
11. Elevated homocysteine level
12. Elevated hs-CRP (C reactive proteins)
13. Obesity
14. Type A personality
15. Stress associated with work, bigotry, racism, illiteracy, poverty and poor economic Status
16. Alcoholism

The percentage breakdown of the incidence of high cholesterol in various ethnic groups is as follows:

Mexican American females: 50.0%
Mexican-American males: 49.9%
White females: 49.7%
White males: 47.9%.
Black males: 44.8%
Black females: 42.1%
American Indians/ Alaska Natives: 26%
Asians/Pacific Islanders: 27.3%

There is a high incidence of elevated cholesterol levels in the United States as shown by the following statistics of those segments of the population with 240g/dl or higher:

16.1% White males
18.2% White females
14.1% Black males
12.5% Black females
16.0% Mexican American males have
14.2% Mexican American females[1]

There is also a high incidence of LDL cholesterol of 130 or higher in the United States as shown by the following statistics: LDL cholesterol is the bad Cholesterol.

Mexican American males: 39.0%
Mexican American females: 30.7%
White females: 33.8%

---

[1]   NHANES [1999-2004] NCHS and NHLBI

White males: 31.7%
Black males: 32.4%
Black females: 29.8% [1]

There are many people in the U.S. with HDL cholesterol less than 40 mg/dL. HDL cholesterol is the good cholesterol. HDL cholesterol helps to take the bad cholesterol out of the body through the stool, thereby decreasing the risk of atherosclerosis in the human body. Conversely, low HDL cholesterol facilitates the development of atherosclerotic diseases in the human body such as:

1. coronary arteries
2. arterial disease of the brain
3. arterial disease of the carotid arteries
4. arterial disease of the abdominal and thoracic aortas
5. arterial disease of the kidneys
6. arterial disease of the peripheral system
7. arterial disease of the male's genital organ
8. arterial disease of the eyes, etc.

The statistical breakdown of the incidence of HDL cholesterol of 40 mg/dL or lower is as follows:

Mexican American males: 27.7%
White Males: 26.2%
Mexican American females: 13.0%
Black Males: 15.5%
White females; 8.8%
Black females; 6.9%[2]

Hyperlipidemia (too much fat in the blood) is, generally speaking, a genetically transmitted disease. If a person's mother or father has too much fat in his or her blood, this trait is likely to be transmitted to his or her children, resulting in hyperlipidemia, which can lead to the development of coronary heart disease resulting in heart attack and possible early death. Hyperlipidemia is categorized as.

1. High blood cholesterol

---

[1]   HANES [1999-2004]. NCHS and NHLBI
[2]   [NHANES] 1999-2004. NCHS and NHLBI

2. High blood triglycerides
3. High low-density lipoprotein
4. Low high-density lipoprotein
5. Cholesterol/HDL ratio which is LDL/HDL greater than 7.13.

A person with LDL/HDL ratio greater than 5.57 is at high risk for developing Cardiovascular Disease (CVD).

Each one of these different components of hyperlipidemia represents an independent risk factor which, when abnormal, results in coronary heart disease

The different types of abnormal lipids that can be found in the blood are:

1. Hyperlipidemia
2. High cholesterol
3. High triglycerides/high cholesterol
4. High low-density cholesterol (LDL)
5. Low high-density cholesterol (HDL)
6. High cholesterol/LDL ratio
7. High VLDL cholesterol

All these abnormal lipids are transmitted genetically from parents to their children to one degree or another. According to a recent report that appeared in the *New England Journal of Medicine*, Vol. 342 No. 12 (March 23, 2000), four new markers of inflammation were found to be predictors of future development of coronary heart disease. These are hs-CRP, serum amyloid A., interleukin-6, and sICAM-1. According to the authors of the report, the hs-CRP was the most sensitive predictor when found to be elevated.

Normal blood cholesterol is from 130 to 200 mg/dl. Normal blood triglycerides are 60–150 mg/dl. Normal HDL is 35–80 mg/dl. Normal LDL is less than 130 mg/dl. In men, normal cholesterol/HDL is less than 3.4. Normal LDL/HDL is less than 2.8. In women, normal cholesterol /HDL ratio is less than 3.27 and the LDL/HDL is less than 2.34.

Most people believe that blood cholesterol level is the only thing that matters when dealing with abnormal fat levels in the blood. This is wrong, because a person may have perfectly normal total blood cholesterol and yet have significant hyperlipidemia, predisposing that person to coronary artery disease. Be aware that the quick cholesterol test may be misleading if normal. Normal blood cholesterol by itself is not enough to tell if a person has abnormal genetically transmitted lipid. There are five basic cholesterols in the blood:

1. Total cholesterol
2. High-density lipoprotein (HDL)
3. LDL cholesterol
4. Triglycerides
5. VLDL (Very low-density lipoprotein)

HDL is the cholesterol that takes the regular cholesterol from the blood, carries it into the bowel and the colon, mixes it with stool, and carries it out of the body. If the HDL is low, less than 45 mg/dl, then there is not enough of it in the blood to complex with bad cholesterol to remove it from the body. This is a genetic abnormality transmitted from parents to children. More appropriately, these lipid abnormalities are called hyperlipoproteinemias. When both the fasting total cholesterol and the LDL are elevated, that is type 2a hypercholesterolemia.

When the fasting total cholesterol, the LDL cholesterol, and the triglycerides are elevated, that is type 2b hypercholesterolemia. When the total cholesterol is high and if the triglycerides are very high, that is type 3 hyperlipidemia. When the triglycerides are very high and the VLDL is high, that is type 4 hyperlipidemia. High chylomicrons, high VLDL, high triglycerides and cholesterol manifest type 5 hyperlipidemia.

Type 1 hyperlipoproteinenemia is manifested by high chylomicrons.

Secondary hyperlipoproteinemia is seen in association with several medical conditions such as diabetes mellitus, hypothyroidism, uremia, and nephrotic syndrome, alcoholism with acute or chronic pancreatitis, ingestion of oral contraceptive, etc.

First, high triglycerides and VLDL may be evident on the skin and under the eyes as deposits (xantomas). Second, VLDL, triglycerides, and high cholesterol may be high in diabetic men who develop ketoacidosis. Third, high triglycerides, high cholesterol, diabetes mellitus, and hypertension may be present persistently in obese men (Syndrome X).

The use of birth control pills or ingestion of any estrogen-containing pills can raise the level of VLDL and triglycerides. One of the dangers of taking estrogen-containing pills is the possibility of high level of lipids. It is important to know the lipid level in a person before he or she starts taking estrogen pills. If a person has an elevated lipid level, estrogen-containing medication may be harmful to his health by increasing the blood lipid, further predisposing that individual to heart attack, stroke, phlebitis, pulmonary embolism, etc.

Alcohol use is also associated with elevated lipids in the blood, such as triglycerides and, in particular, high very-low-density lipoprotein and chylomicrons. Type 5 hyperlipidemia and sometimes Type 4 hyperlipidemia may be associated with increased alcohol use. Type 5 hyperlipidemia may cause acute pancreatitis, which is a serious medical condition and if left untreated can be fatal.

Hyperlipidemia causes coronary artery disease because in a high lipid state, lipid is deposited within the inside of coronary arteries, causing gradual narrowing of these vessels and resulting in coronary occlusive heart disease. When the vessels around the heart are narrowed, the condition called angina pectoris frequently develops. Angina pectoris is manifested by chest pain caused by the lack of oxygen being delivered to the heart muscle. The pain occurs when tissue is deprived of oxygen causing a series of substances, called kinins, to be secreted in and around that tissue which then develops into a burning pain. A good example of kinins is the substance that develops in a blister in one's finger or toe. If one bursts the blister right away, the liquid that forms within it causes a burning sensation because that liquid contains kinins.

High lipoprotein-a is also associated with coronary heart disease, myocardial infarction, stroke, Transient Ischemic Attack (TIA) and Deep Vein Thrombophlebitis (DVT).

A high level of homocysteine level is also associated with coronary heart disease, stroke, myocardial infarction, TIA, and DVT. Both these conditions are genetically transmitted, and can cause thrombosis to occur anywhere in the body, both in the high flow and the low flow systems.

Diet plays a major role in the prevention of obesity and the prevention and control of hypertension. Diet also plays a major role in both preventing and controlling the levels of cholesterol and triglycerides in the blood. The so-called soul food that blacks and other people of color like to eat so much is a legacy of slavery some 500 hundred years ago.

However, "soul foods" have too much fat, carbohydrates, and salt, and are too spicy. These foods taste good, but they are unhealthy. Therefore, it is fine to eat them every now and then; but when a person eats them on a daily basis, it increases his or her chances of becoming obese and raising his or high blood pressure and his or her cholesterol.

A combination of obesity, high blood pressure, and high level of fat in the blood is responsible in part for the high incidence of coronary artery disease, stroke, and deaths in the United States.

To prevent these things from happening, popular diets must be modified. Diet is very ethnic in its origin. People of different ethnic backgrounds have different tastes for different foods, and that is fine, except that one has to understand that everything has to be done in moderation. If a person eats fat and salt-laden foods too often, that person is likely to pay the consequences with an increased incidence of high cholesterol, diabetes, obesity, coronary artery disease, hypertension, and stroke. Many blacks and other minorities, in large measure, suffer from these conditions because of poor living conditions, poor diet, and overall poor economic conditions.

The billions of dollars spent by fast food restaurants to advertise their foods entice blacks and the poor to feast on different fast foods even though these foods are extremely unhealthy.

Because of all of these factors, a high level of disparity exists in the incidence of high cholesterol and other abnormal lipids in Blacks, Hispanics, Native Americans, Alaskan Natives, Asians/ Pacific Islanders, and poor people of all ethnic backgrounds.

All of this is made worse by the present economic situation in the U.S. which is forcing many people who previously had good economic standing and relatively healthy diets, to start eating unhealthy foods to feed themselves and their families with now severely constricted household budgets. The current economic crisis has increased the pool of people who are obese, diabetic, hypertensive, have high cholesterol, and are at higher risk for strokes, heart attacks and death. In effect, this segment of the population has now joined people of color who have for so many years suffered from the effects of unhealthy diets.

An understanding of these issues and doing the things that are necessary to modify them will go a long way to prevent or at least decrease the high incidence of hyperlipidemia / high cholesterol and all their accompanying diseases.

CHAPTER 6

DISPARITY IN OBESITY IN THE UNITED STATES

Obesity is a serious medical problem. Morbid obesity is a frequent cause of disability in men, women, children, and increasingly more frequent in children and adolescents.

According to WHO, by 2015 there will be 2.3 billion overweight people in the world and 700 million individuals will be obese. In 2005, there were about 20 million obese children in the world, according to the WHO.

Obesity is associated with 440,000 deaths annually in the U.S.

The low basal metabolism gene is at the core of the condition of obesity/overweight. Low basal metabolism is passed from parents to offspring. Genetic traits are adaptable, penetrating, and "transmissible". Obese people, the world over, inherited the obesity gene from their African ancestors. This gene is disseminated among people of all ethnic groups. The human race and all is its original DNA genetic traits began in Africa around 60,000 years ago.[1] Obesity/overweight is therefore a genetically transmitted disease.

Obesity, when it is not associated with malfunction of the endocrine system, is always the result of eating too much of the wrong foods; foods which are too rich in fat, salt, and simple carbohydrates and too low in protein. The most effective diet is a diet that is low in simple carbohydrates, fats, and high in protein.[2]

---

[1] *The* Journey *of man: A Genetic Odyssey, Princeton University Press(2002).*

[2] Alcena, Valiere, M.D., F.A.C.P., *The Third World Tropical Diet Health Maintenance and Medical Management Program*, Alcena Medical Communication Inc (1992).

In 2005, 142,000,000 adults were obese or overweight in the U.S. Of this number, there were 73,000,000 males and 69,000,000 females. This represented 66 percent of the adult population in the U.S. In 2008, thirty eight percent of children and adolescents were obese in the US.

Presently, two-thirds of the U.S. population or 306, 637,411 are obese/overweight. Of these, 35.5% of men are obese and 32.2% of women are obese. However among blacks, the statistics are quite different. Obesity among black men is slightly lower as compared to whites. On the other hand, obesity is 80% higher in black women as compared to white women. In fact, obesity is more common in African American women than any racial group in the U.S. In addition, there exists a clear disparity in the incidence of obesity/overweight in Blacks, Hispanics, and Whites in the US.

The root cause of the disparity that exists in obesity/ overweight among minorities in America is primarily racism and the poverty that is associated with this major Psychological/ Psychiatric disease. Most minorities begin to experience the effects of racial discrimination from the time they are in their mother's wombs because sick and malnourished mothers are more likely to give birth to sick and malnourished infants.

Racism in America may be described by the following formula: racism = poverty+ malnutrition + poor education + low income + poor living conditions + stress + poor health + a life full of pain both mental and physical.

Unfortunately, many minorities and other poor individuals are forced to eat foods that are known to cause obesity/overweight. Such foods include fast foods, so-called "soul food", foods that contain too much fat, foods with simple carbohydrates with too much salt and too little protein, and not enough fruits and vegetables.

Exercise has an important role to play in the fight against the obesity/overweight syndrome. However, too many members of minority groups and other poor individuals find themselves with insufficient time to go to the gym to exercise because they are too busy working two to three jobs just to survive. Required gym fees are simply not in the budget. However, in fact, one does not need to join a gym to exercise. Simple activities like walking, running, swimming and bicycling can lead to calorie loss and consequently better health.

In the USA, many factors interplay in causing the high incidence of obesity. For instance, the dietary industry spends somewhere from 50 billion dollars a year selling their different products and varying dietary programs. The food industry spends somewhere around 36 billion dollars a year advertising the different food products and agricultural materials they produce to encourage greater consumption. At the same time, the medical profession devotes very little time and resources in the prevention and treatment of obesity.

The main reason that the medical profession in the United States spends such precious little time in the prevention and management of obesity is that insurance companies could not care less about obesity and will not pay physicians to provide medical care for obese patients or for the measures necessary for preventing obesity in the first place. The federal and state governments are not doing very much either because the government, in general, spends somewhere around 50 thousand dollars a year on nutritional and other educational programs addressing obesity - a pitiful gesture at best.

In the U.S., about 132 billion dollars are spent every year treating obesity and its different complications which include:

1. Cancer of various types, including colon, prostate and pancreatic cancer;
2. Heart disease
3. Adult-onset diabetes
4. Gall Stones
5. Cholecystitis
6. Pancreatitis secondary to Gall Stones
7. Hypertension
8. Stroke
9. High cholesterol
10. Deep Vein Thrombophlebitis
11. Pulmonary embolism secondary to deep vein thrombophlebitis 15. Sleep Apnea, etc., etc.

Notably, people who live in the underdeveloped world who, by necessity, are forced to eat a meager diet are less obese than people who live in the developed world such as in the United

States. In the underdeveloped world, most people eat plenty of fresh fruits, green vegetables, grains, yams, plantains, bananas, less red meat, and plenty of fish.

Typically, people in the so called underdeveloped or "third world" exercise more because they often have to walk long distances to the farm or the marketplace and they walk to the river to fetch water, etc. Sometimes, long hours are spent working under the hot sun on farms. Elsewhere, long hours are spent working in sweatshops. Still others in different parts of the world work at home doing all sorts of chores around their homes. All these activities result in burning up calories which is quite important in maintaining good body weight.

A combination of such factors leads to the incidence of less obesity in the underdeveloped world. Nevertheless, people in the underdeveloped world still carry the obesity gene and are able to pass it on even though they themselves are able to work off the extra fat that is deposited into their tissues as predetermined by their hereditary trait.

In addition, it must be understood that if a person is fat, he or she is likely to give birth to a fat baby, even though the baby may not be fat at birth. Because of the low basal metabolism gene, the baby will likely grow up to become a fat adolescent and a fat adult.

The foods that people in the underdeveloped world typically eat are high in protein, low in fat, high in vitamins and fibers, and low in simple carbohydrates. This combination of foods contains high complex carbohydrates. Examples of high complex carbohydrates foods include: rice (brown), bagels, pretzels, pasta, yams, bananas, plantains, potatoes, dumplings, corn, cereals, breads (whole wheat), whole grains, tortillas, waffle, grits, millet, oats, wheat germ, granola, cornmeal, shredded wheat, flour, etc.

High complex carbohydrates are broken down very slowly in the body and provide a lower but longer level of energy. This is what makes them ideal food products, in that a person can eat high complex carbohydrates to satisfy hunger and, at the same time, provide vitamins and fiber for regular gastrointestinal functioning, especially important for proper bowel movements. Therefore, complex carbohydrates do not contribute to an increased caloric level, which can result in obesity. On the other hand, simple carbohydrates, such as sugar-containing foods, can be broken down in the liver and some of them distributed into the tissues and muscles, resulting in obesity when consumed in large quantities.

The above described foods, when prepared in vegetable oil, are either boiled or broiled and not fried, satisfy hunger, provide needed vitamins such as Vitamin A, Vitamin K, the B Vitamins, including B6, B12, etc. All of these are important nutrients for the body.

Obesity and Atherosclerotic Disease

Obesity is also associated with Atherosclerotic disease in the body. The persistent high level of insulin that is present in the bloodstream of the obese person causes plaques to develop within vessels throughout his or her body, including the coronary arteries around the heart. When these arteries are occluded, blood flow is impeded, preventing proper oxygen delivery to the muscles of the heart and causing pain in the chest to occur.

Because there are plaques in the coronary arteries, sudden closure of one or several of these coronary arteries can result in a heart attack and, frequently, death of the obese person. Plaques inside arteries can develop in all organs in the human body. This development of plaque in the human body can cause a long list of diseases that can either kill or render one disabled to one degree or another.

In summary, the drain on the pocketbook from monies spent on so-called diet programs is enormous, often for those who are least able to afford it. These programs are expensive and, according to the U.S. Government, have questionable motives, not necessarily including true weight loss and, more importantly, the maintenance of a proper body weight. In fact, some of these diet programs may actually be medically dangerous if entered into without proper medical supervision.

Every year, 132 billion dollars are spent in the U.S. to treat obesity and its multitude of associated medical problems.[1]

Learning a new way to prepare foods and a change in eating habits will go a long way in helping obese people to lose weight without participating in expensive and potentially

---

[1]   MMWR, Vol. No. 36, Sept. 13, 2002 (DC/NCHS).

dangerous dietary gimmicks designed primarily to make money for the people who advocate such programs.

Paying close attention to the basics and adhering closely to a healthy lifestyle of exercise, good diet, low alcohol consumption, as well as frequent visits to the physician's office for proper health screening will go a long way towards decreasing the incidence of obesity and its associated diseases with their often devastating consequences.

CHAPTER 7

DISPARITY IN DIABETES MELLITUS IN THE U.S.

Five percent of the world populations of 6,720,517,952 or 336,025,898 individuals worldwide have diabetes mellitus type II according to the Centers for Disease Control (CDC).

In Europe, 8% of the adult population is diabetic, and 60,000,000 of the adults are pre-diabetic according to the International Diabetes Federation.

There are now 306, 637,411 people in the U.S. Of this number, there are 24,000,000 diabetics. Ninety to ninety-five percent of diabetics in the U.S have type 2 diabetes and five to ten percent have type 1 diabetes. Eight per cent of the U.S. population has diabetes and 54, 000,000 individuals are pre-diabetics most of whom do not know they have the disease, according to the CDC.

About 1.7 million individuals in the U.S. have type 1 diabetes. The percentage breakdown among the various ethnic groups is as follows:

Mexican American males: 13.9%
American Indians/Alaska Natives: 16%
Black males: 11.6% Hispanic males: 9.4%
White males: 9.2%
Asian males: 6.3%

The incidence of diabetes is 2-4 times higher in African American women, Hispanic women, American Indian women, and Asian/Pacific Islander women as compared to White women.

The cost of the over all treatment of diabetes was $174 billion in 2007.

Gestational diabetes mellitus (GDM) represents the third most common form of diabetes. Most of the time, women who develop GDM remain free of diabetes after delivering their babies. However, 20-50% of the time, women who had GDM has the possibility of developing diabetes mellitus within five to ten years.

There also exist clear ethnic disparities among Black women, Hispanic women and foreign born women compared with White women in the U.S. in regards to which groups of women are more likely to develop gestational diabetes mellitus. White women have a lesser likelihood of developing GDM as compared to minority women.[1] The reasons for this disparity include the fact that minority women face a much more difficult daily lives and struggle than do white women. These daily life problems are associated with such things as poverty, poor diet, obesity, and stress, all of which can lead to diabetes pre- pregnancy and during pregnancy.

Diabetes mellitus can be prevented according to the DREAM study, (Diabetes Reduction Assessment with ramipril and resiglitazone Medication) - if individuals at high risk for diabetes are treated with Rosiglitazone (Avandia). Source: Lancet

Avandia works by making fat cells more sensitive to insulin thereby preventing the blood sugar from being elevated in the blood stream and decreasing the level of insulin in the blood.

Diabetes mellitus has increased 600% in the United States since 1958 and it is estimated that the incidence of diabetes mellitus will rise by 35% in the next ten years. The worldwide incidence of diabetes mellitus was 171 million in the year 2000 and it is estimated that in the year 2030, the worldwide incidence of diabetes mellitus will be 366 million. Presently there are 230 million people in the world with diabetes.

There are 57 million people with in the U.S. with pre-diabetes.

Normal blood glucose is from 65mg/dL to about 109 mg/dL. If the fasting blood sugar is 110 -124 mg/dL, this is pre-

---

[1]   Am International Journal of Obstetrics and Gynecology, Volume 115, Issue 8, pp 969-978 (published on line: 28 June 2008)

diabetes. If the fasting blood sugar is 125 mg/dL or higher, this is diabetes mellitus.

The newer way of testing for diabetes is to use the Hemoglobin A1c. If the screening hemoglobin A1C is 5-6%, diabetes is not present. However, if the hemoglobin A1C is 6.0% -6.5%%, prediabetes is probably present. If the hemoglobin A1C is 6.5% or higher, then full-blown diabetes is present.[1]

Diabetes is a very complicated and complex disease that affects all organs in the human body. However, the organs that suffer most from the devastation of diabetes are the so-called end organs consisting of the eyes, heart, kidneys, brain, the peripheral vascular system, the peripheral nervous system.

Other organ systems that are frequently affected by diabetes mellitus causing severe pain and suffering are the nervous system, causing peripheral neuropathy with pain, numbness, and coldness in the toes, feet, and fingers. If severe enough, diabetic neuropathy can cause the affected person to be unable to walk. The skin is one of the most frequently affected organs in patients with diabetes. Diabetes affects the colon by causing constipation. Diabetes affects the stomach by causing gastroparesis with frequent indigestion, bloating, and burning in the stomach. Diabetes affects the urinary system by causing urinary retention.

The eyes are affected by diabetes because of the series of damages that elevated blood sugar causes inside the eyes. These damages result in bleeding within the eyes, a condition called diabetic retinopathy. Diabetic retinopathy is a condition that, if left untreated, can lead to blindness. The treatment for diabetic retinopathy is laser surgery.

The heart is affected by diabetes by causing hardening of the arteries, known as atherosclerosis. Atherosclerosis causes narrowing of the coronary arteries, resulting in ischemic heart disease, which causes angina pectoris and frequently results in myocardial infarction (a heart attack).

The kidneys are affected by diabetes through damage to the kidney tubules and glomeruli, resulting in diabetic nephropathy of different degrees. Diabetic nephropathy can cause protein

---

[1]   Internal Medicine News, Vol. 42, No 12, June 15, 2009

loss, microalbuminuria and, ultimately, nephrotic syndrome. The result of that constellation of abnormalities is elevated serum BUN, creatinine, potassium, and renal insufficiency, which usually results in end-stage renal failure. End-stage renal failure is treated with dialysis.

The brain is affected by diabetes by way of atherosclerosis of the arteries inside the brain, causing narrowing of these vessels, and preventing easy flow of blood and oxygen, which can result in a stroke.

Some of the late signs and symptoms of diabetes are.

1. Blindness
2. Chronic kidney failure
3. Coronary artery disease
4. Recurrent leg and feet ulcers with frequent loss of lower limbs
5. Peripheral neuropathy
6. Sexual impotence
7. Loss of libido
8. Gastroparesis.
9. Infertility
10. Constipation
11. Recurrent fungal infections of the skin, sinuses, etc.

Another frequent problem that develops from diabetes is urinary tract infection because diabetes damages the smooth muscle and nerves within the bladder, causing poor contraction of the bladder and preventing complete excretion of urine. The residual urine that stays in the bladder serves as a culture medium allowing for bacterial growth, and the result is recurrent urinary tract infection and all its many potential complications.

Some dietary foods, such as those rich in fat and simple sugars , play a major role in people being overweight and insulin-resistant. In poor people and people of color, poverty plays a major role in their inability to afford better foods. Therefore, they eat the foods they can afford. The types of foods that they can afford are frequently of poor quality. Even when the foods are of very good quality, the way in which they are prepared makes it too rich in fat and carbohydrates.

Poor quality foods are compensated for by preparing them in a way that makes them more palatable by sometimes curing them with a fruit juice called "sour". This fruit is very juicy with a flavor of "bitter orange". When plenty of salt, hot pepper and other spices are added, the food is more palatable and its taste improves but not necessarily the quality. This is the legacy of the so-called soul foods, which are frequently eaten by many people of color. Soul foods are detrimental to good health. Whites and Asians eat these foods less frequently, and when they do, they eat them as curiosities.

Fast foods like hamburgers, cheeseburgers, hot dogs, fried chicken, spareribs, pizza, and, soft drinks etc., have proliferated in U.S. society and, are easy to purchase and cost less. Young people of all ethnic backgrounds eat these foods and drink these drinks very often, and, are getting fat, with many of them developing type II diabetes, heart disease, cancer, and other serious diseases at an alarming rate.

These foods, although popular, are definitely not very nutritious and certainly not particularly healthy. It is perfectly fine to eat fast foods if done infrequently, but if one makes it a habit of feasting on these foods on a regular basis, then the health consequences can be dire indeed.

Obesity and diabetes are intertwined, and as just outlined, they are both genetically transmitted diseases and interact together. When they interact together in the same person, it makes it much more difficult to provide medical care.

Eating a diet that is rich in fruits, vegetables, protein, high in complex carbohydrates, and low in fat, simple sugar, and salt, exercising regularly to burn calories, will all contribute to weight loss and increasing insulin-sensitivity, which in turn decreases blood sugar and the incidence of diabetes.

The disparity that exists in the incidence of diabetes in blacks and other minorities as compared to whites is glaring and must be reversed. Diabetes is the leading cause of end stage renal failure leading to chronic dialysis in blacks.

Diabetes mellitus, while not a curable disease, is definitely a treatable disease. There are plans underway for pancreatic transplants and if these become successful, then the disease can, at that point, be considered curable. Insulin pumps are also already in use. These pumps add a great deal to the treatments of diabetics requiring insulin.

There is now research underway to try to determine the cause of Type I diabetes and the hope is that someday, the answer will be found. Meanwhile, it is important for diabetics to learn as much about diabetes mellitus as they can, and in the case of diabetics who are obese, it is important that every effort be made to get the excess weight under control and so help better control their diabetes.

It is important to note that deaths from heart disease in individuals with diabetes mellitus are about four times higher than the death rates from individuals not suffering from diabetes.

CHAPTER 8

DISPARITY IN HEART DISEASEIN THE U.S.

Heart disease is the leading cause of death among people of all ethnic groups in the U.S.

There are 80,700,000 individuals with cardiovascular heart disease (CVD) in the U.S. One in three people in the U.S. have one type or another of CVD.

Cardiovascular disease is made up Hypertension, Coronary Artery Disease (CAD) and Stroke.

In the U.S., 73 million people have hypertension, 16 million have CAD, and 5,300,000 people suffer strokes.

8,100,000 people have myocardial infarction. 9,100,000 have angina pectoris, 6,600,000 have congestive heart failure and 650,000-1,300,000 have congenital heart disease.

In the white population, 12.0 percent have heart disease, 6.6 percent have CAD, 21percent have hypertension, and 2.3 percent have had a stroke.

In blacks, 10.2 percent have heart disease, 6.2 have CAD, 31.2 have hypertension, and 3.4 percent have had a stroke.

Among Hispanics/Latinos, 8.3 percent have heart disease, 5.9 percent have CAD, and 20.3 percent have had a stroke.

In the Asian population, 6.7 have heart disease, 3.8 percent have CAD, 19.4 percent have hypertension, and 2.0 percent have had a stroke.

Among Native Hawaiians/Pacific Islanders, 22.4 percent have hypertension.

Among American Indians/Alaska Natives, 13 percent have heart disease, 2.5 percent have CAD, 25.5 percent have hypertension, and 5.8 percent have had a stroke.

There are 32.500.000 males with cardiovascular disease in the U.S. representing 34.4% of all men. In the white male population, 34.3% have cardiovascular disease, in black males, 41.1% have cardiovascular disease, and among Mexican American males, 29.2% have cardiovascular disease. [1]

Every year, 1,357,000 people die of CAD in the U.S. This means that about 2,400 Americans die of CVD every day in the U.S.

There are 6,600,000 individuals who have congestive heart failure in the U.S. and there are 500,000 new cases of congestive heart failure every year. 2,400,000 people suffer from congestive heart failure in the US.[2]

Every year, there are 1.1 million heart attacks in the U.S. and 450,000 of them are recurrent heart attacks.[3] Every day, 2,600 people die of cardiovascular disease in the U.S. which represents on an average one death every 33 seconds.

In the year 2000, there were 2,400,000 deaths in the U.S. from different causes and 1,415,000 of these deaths were due to cardiovascular disease of different types. Each year over 500,000 people die of coronary heart disease in the U.S. That is to say, cardiovascular heart disease is the number one killer in the U.S.

The total number of deaths due to cardiovascular disease in the year 2002 was 927,448. It is estimated that there were 7,100,000 males with coronary heart disease in the U.S. in the year 2002, and during that same period, 4,100,000 of these men have had myocardial infarctions.

In **2004, 451,326** people died of coronary heart disease.

Of the people who died of coronary heart disease, **233,538** were men and 217,288 were women. A full 25% of people die within a year after suffering a heart attack. One-half of individuals

---

[1]

[2] National Health and Nutrition Examination Survey NHANES (1999-2002)

[3] *Morbidity and Mortality: 2002 Chart Book on Cardiovascular, Lung and Blood Diseases* Bethesda, Maryland National Heart, Lung and Blood Institute, May 2002.

under age 65 who have had a heart attack die within eight years after suffering the first heart attack from coronary heart disease. One-half of all people who died suddenly due to coronary heart disease had no previous symptoms of the disease.[1]

The direct and indirect cost of coronary heart disease is projected to be 151.6 billion dollars in 2007.

Blacks and other people of color are less likely to be offered cardiac catheterization to evaluate them for the likelihood of coronary artery disease. In addition, when people of color are found to have coronary occlusive disease, they are less likely to be offered coronary bypass to treat their coronary artery disease. This is a contributing factor to the disparity in the survival rate of people of color and other segments of the population.

People of color live under conditions that are oftentimes more stressful than most whites, including factors such as supporting and raising a family on a meager income and working two jobs to try to pay bills. While poor whites face some of these same economic problems, they do not have to deal with the constant indignities of daily racial discrimination and harassment of different types. These multitudes of responsibilities and the psycho-socio-economic problems create an unhealthy situation which predisposes blacks and other people of color to constant stress which, in turn, creates a perfect formula for the development of coronary heart disease, heart attacks and sudden death.

Constant stress causes secretion of a high level of adrenalin which causes the blood pressure to rise. High blood pressure, in turn, can lead to coronary artery disease and heart attack and death. Also, the high level of adrenalin over-stimulates the heart resulting in cardiac arrhythmias which can result in sudden death. This, in part, explains why the rate of death from heart disease is so much higher in blacks and other minorities as compared to whites.

The combination of obesity, hypertension, diabetes mellitus, hyperlipidemia and insulin resistance, referred to as syndrome X or metabolic syndrome plays a major role in the causation of coronary artery disease.

As has already been frequently noted, the diet of most people of color is less healthy than that of whites, because in part the economic situation of that subgroup is poorer and creates

[1] Heart Disease and Stroke Statistics: 2005 Update American Heart Association

a lifestyle that predisposes them to poorer cardiac health. As already described, the diet of this subgroup is too rich in fats, carbohydrates, and salt and too poor in protein and fibers. Such a diet is guaranteed to result in many serious medical problems including heart disease. The cholesterol level of people of color is higher than that of whites. As has already been seen, obesity is also more prevalent in people of color than in whites.

The incidence of hypertension is higher in people of color than in whites. Stress, as a psychological condition, is more prevalent in people of color than in whites. The incidence of cigarette smoking is higher in people of color than in whites, percentage-wise. The incidence of alcohol consumption in the U.S. percentage-wise, is higher in people of color than in whites. The percentage of IV drug U.S. is higher in people of color than in whites.

All these factors contribute in one form or another to overall poor cardiac health. The disparity in the occurrence of these negative factors results in the overall rate of heart disease morbidity/mortality being significantly higher in people of color.

This is compounded by the fact that when people of color present to the emergency room with symptoms of heart disease, they are much less likely to get proper medical care as compared to white patients who present with the same or similar symptoms. The reason for the dissimilar treatment in the emergency room is that pain being experienced by persons of color, and frequently women, is usually attributed to other factors. The result is that they are less likely to be assigned to a coronary care unit and they are less likely to be offered cardiac catheterization, angioplasty, or coronary bypass.

Those who receive the fastest attention and receive the highest priority in the health care system in the U.S. are white males. This is so, in part, because more often than not, the physicians making the decisions as to who gets what type care are white male doctors in training (interns, residents, and fellows). These young physicians work at the front line in the emergency rooms inside the hospitals. In community hospitals where there are no training programs, the attending physicians see their own patients and control the quality of the care given.

The availability of health insurance plays a major role in the **type of care that is offered when individuals present seeking health care. For instance, according to a recent report, 46** million individuals do not have health insurance in the U.S. and many of them are poor minority men and women. White males and white females have better access to health care than black males and black females. White males live longer than do black males and White females live longer than do Black females. The median survival for white males is 76 years and the median survival of black American males is 70 years, a difference of 6 years in favor of white American males. The median survival for White females is 81years and median survival for Black females is 77 years, a difference of 4 years in favor of White females in the U.S.

The bottom line is that taken together, white American males live six years longer than do black American males and white females live four years longer than do black females. This is not due to genetic inheritance but rather to a multitude of economic, educational, professional, racial, and psychosocial advantages that white men and white women enjoy over black men and women in the U.S.

Poor people and most people of color tend to go to physicians with diseases of all sorts when these diseases are already in their advanced stages. They frequently ignore their symptoms and go to physicians when the medical problems are frequently more challenging to deal with.

The incidence of heart disease and the incidence of death from heart disease can be brought down significantly if the gravity of the disease is more clearly understood and the necessary preventive measures – most notably, a change in lifestyle - are taken.

Another point that needs to be made is that obesity is frequently associated with diabetes, hypertension, and hyperlipidemia and all four conditions represent high risk factors for the development of coronary heart disease. Diabetics frequently have atypical symptoms of coronary heart disease and they frequently have silent heart attacks. Understanding these facts and paying close attention to them will decrease the incidence of death from heart disease.

CHAPTER 9

DISPARITY IN CANCER IN THE U.S.

In the year 2008, about 1,437,180 individuals were diagnosed with cancer in the United States. Of that total, 745,180 were in men and 692,000 were in women. In addition, more than 1 million individuals develop basal cell carcinoma of the skin yearly. In 2008, according to the American Cancer Society, 565,650 individuals died of cancer in the United States and of that number, 294,120 were men and 271, 530 were women. This means that more than 1,500 individuals die of cancer everyday in the U.S. One in every four deaths in the United States is due to cancer. In 2007, there were 10.5 million cancer survivors in the U.S. Since 1990, roughly 5 million people have died of cancer of different types in the U.S. The combined five-year survival rate for all cancers diagnosed and treated in the United States is 62%; the five-year survival rate is higher for some cancers and lower for other cancers.

The estimated new cancer cases in the U.S. in men in 2008 were:

Prostate: 25%
Lung & bronchus: 15%
Colorectal: 10%
Urinary bladder: 7%
Non-Hodgkin lymphoma: 5%
Melanoma: 5%
Kidney: 4%
Oral cavity: 3%
Leukemia: 3%
Pancreas: 3%
All other sites: 20%

In women, the new cancer cases in 2008 were:

Breast: 26%
Lung and bronchus: 14%
Colorectal: 10%
Uterus: 6%
Non-Hodgkin lymphoma: 4%
Thyroid: 4%
Melanoma: 4%
Ovary: 3%
Kidney: 3%
Leukemia: 3%
All other sites: 23% [1]

American Cancer Society, 2008

The number of deaths in men from cancer in the U.S. in 2008 was:

Lung and bronchus: 31%
Prostate: 10%
Colorectal: 8%
Pancreas: 6%
Liver & intrahepatic bile ducts: 4%
Leukemia: 4%
Esophagus: 4%
Urinary bladder: 3%
Non-Hodgkin lymphoma: 3%
Kidney: 3%
All other sites: 24%

The number of deaths in women from cancer in the U.S. in 2008 was:

Lung & bronchus: 26%
Breast: 15%
Colorectal: 9%
Pancreas: 6%
Ovary: 6%
Non-Hodgkin lymphoma: 3%
Leukemia: 3%

Uterus: 3%
Liver & intrahepatic bile duct: 2%
Brain/ONS: 2%
All other sites: 25%[1]

The disparity that exists both in the development of cancer and deaths from cancer in all sites in the human body in blacks vs. whites in the U.S. is quite glaring.

	Black Women	White Women	Difference
All sites	189.3	161.4	1.2
Myeloma	6.3	2.8	2.3
Stomach	5.8	2.6	2.2
Uterine cervix	4.9	2.3	2.1
Esophagus	3.0	1.7	1.8
Uterus	7.1	3.9	1.8
Small intestine	0.5	0.3	1.7
Larynx	0.8	0.5	1.6
Pancreas	12.4	9.0	1.4
Colorectal	22.9	15.9	1.4
Liver and intrahepatic ducts	3.9	2.8	1.4
Breast	33.8	25.0	1.4
Gallbladder	1.0	0.8	1.3
Urinary bladder	2.8	2.3	1.2
Oral cavity and pharynx	1.7	1.5	1.1

Per 100,000[2]

	Black Men	White Men	Difference
All sites	321.8	234.7	1.4
Prostate	62.3	25.3	2.4
Larynx	5.0	2.2	2.3
Stomach	11.9	5.2	2.3
Myeloma	8.5	4.4	1.9
Oral cavity and pharynx	6.8	3.8	1.8
Small intestine	0.7	0.4	1.8
Liver and intrahepatic bile ducts	10.0	6.5	1.5

---

[1] American Cancer Society, 2008
[2] National Cancer Institute, 2007

Colorectal	32.7	22.9	1.4
Esophagus	10.2	7.7	1.3
Lung and bronchus	95.8	72.6	1.3
Pancreas	15.5	12.0	1.3

Per 100,000

Overall, according to figures published by the American Cancer Society, Blacks are about 34% more likely to die of cancer than Whites are in the United States. That is an astonishingly great disparity.

The percentage of blacks who die of cancer in terms of five year survival is much higher than the percentage of whites who die of cancer within the same time span.

5 YEAR CANCER SURVIVAL RATES IN THE U.S.:

	Whites	Blacks	Difference
All sites	67	57	10
Breast (females)	90	78	12
Colon	66	55	11
Esophagus	18	11	7
Leukemia	51	40	11
Non-Hodgkin lymphoma	65	56	9
Oral cavity	62	41	21
Prostate	99	95	4
Rectum	66	58	8
Urinary bladder	81	65	16
Cervix	74	66	8
Uterus	86	61	21

Cancer is the second leading cause of death in the United States next to cardiovascular heart disease. Cancer is also a very costly disease. The overall direct and indirect cost of cancer in the U.S. in 2007 was 219.2 billion dollars according to the American Cancer Society.

What is cancer?

Cancer develops when a cell loses its ability to grow and multiply in a normal growth pattern. A good example of that is contact inhibition. When a normal cell is placed in contact with a hard surface in a Petri dish, the normal cell stops growing.

However, in the case of an abnormal cell, it continues to grow because it has lost its contact inhibition ability which allows it to grow uncontrollably, developing into a cancer growth.

The cancer cells fail in the process of cell-to-cell interactions. The development of cancer is a multi-step and multi-factorial process. In the multi-step and multi-factorial processes, there is normally a balance between growth-promoting genes (proteins) and growth-suppressive genes (proteins). Once mutation occurs for one reason or another, the growth-promoting genes (oncogenes) stop the suppressive effects of the suppressor genes. The growth-promoting genes take control and promote abnormal cell growth, resulting in the formation of a cancerous clone of cells, resulting in what is known as cancer. This is an out-of-control process of cell growth which ultimate goal, is to take over the body in which it is growing and destroy it.

The first step in the genesis of cancer is the process of oncogene. Oncogenes can be brought about by a hereditary or familial transmission of a protein from parent or parents to the fetus at the time of conception. That protein (oncogene or oncogenes) can then enter over many years into a multi-step process, interactions, and reactions that can cause a cell or group of cells to mutate. Once that mutation occurs, then the cell or group of cells loses their ability to grow and multiply normally. The abnormal growth of cells then becomes a cancer mass. There are many causal agents that can result in damage to the cells that causes the mutation to occur.

All of the following can damage the DNA/RNA materials inside a cell resulting in a malignant mutant:

1. Transmission of a hereditary cancer oncogene ;
2. Exposure to oncogenic viruses such as Epstein bar virus, which can cause nasopharyngeal carcinoma ;
3. Exposure to human papilloma virus, causing cervical cancer;
4. Exposure to either hepatitis B or C virus, causing cancer of the liver;
5. Exposure to HTLV-I and HTLV-II, causing T cell leukemia/ lymphoma;
6. Sun exposure causing basal cell carcinoma of the skin;

7. Exposure to carcinogens such as tobacco smoking, causing lung cancer, cancer of the mouth, throat, head and neck, etc.;
8. Exposure to ionizing radiation, causing leukemia, lymphoma and other cancers ;
9. Exposure to toxic chemicals such as benzene, etc., causing malignancies of different types;
10. Consumption of excessive alcohol, resulting in cancer of the mouth, throat and esophagus;
11. Exposure to estrogen, causing increased incidence of breast cancer and uterine cancer in men;
12. Consumption of too much red meats, resulting in increased incidence of breast, uterine and colon cancer;
13. Alcohol consumption and tobacco smoking, associated with increase incidence of cancer of the esophagus;
14. Long-term exposure to toxic pollutants and chemical solvents in the work place, resulting in the development of different types of cancer;
15. HIV-I and II causing AIDS with its high propensity to cause lymphoma and Kaposi's sarcoma;
16. Non-acquired immunodeficiency and its propensity to cause malignancy of different types, etc;

The following is a list of the incidence of the different types of cancers in the U. S. in 2008.

1. Prostate cancer: 186,320
2. Lung cancer: 215,020
3. Colorectal cancer: 148,810
4. Urinary bladder: 68,810
5. Melanoma of the skin: 62,480
6. Cancer of mouth and throat: 60,620
8. Non-Hodgkin/ lymphomas: 66,120
9. Cancer of the pancreas: 37,680
10. Cancer of the kidney and renal pelvis: 54,390
11. Cancer of soft tissue: 10,390
12. Thyroid cancer: 37,340

13. Skin cancer (Non Melanoma): more than 1000,000
14. Esophagus: 16,470
15. Stomach: 21,500
16. Cancer of the liver: 21,370
17. Cancer of the brain and nervous system: 21,810
18. Multiple myeloma: 19,920
19. Cancer of the small intestine: 6,110
20. Cancer of the gall bladder: 9,520
21. Hodgkin lymphoma: 8,220
22. Chronic lymphocytic leukemia: 15,110
23. Chronic myelogenous leukemia: 4,830
24. Acute myelogenous leukemia: 13,290
25. Acute Lymphocytic leukemia: 5,430
26. Testicular cancer: 8,090 27. Cancer of the penis: 1,250[1]

In 2008, the incidence of cancer in women in the U.S. was:

Oral Cavity and Pharynx: 10,000
Esophagus: 3,500
Stomach: 8,310
Small intestine: 2,910
Colorectal: 71,560
Liver: 6,180
Gallbladder: 6,020
Pancreas: 18,910
Lung: 100,330
Soft tissue: 4,670
Skin (not including basal cell and squamous cell): 29,570
Melanoma: 27,530
Breast: 182,460
Urinary Bladder: 17,580
Kidney: 21,260
Brain: 10,030
Thyroid: 28,410
Hodgkin lymphoma: 3,820
Non-Hodgkin lymphoma: 30,670

---

[1] American Cancer Society Cancer Facts and Figures, 2008

Myeloma: 8,730
Chronic Lymphocytic leukemia: 6,360
Acute myelocytic leukemia: 6,090
Chronic myelocytic leukemia: 2,030
Acute lymphocytic leukemia: 2,210[1]

Overall the incidence of cancer in the U.S. in 2008 was 1,437,180.

The incidence of deaths from cancer in the U.S. from all sites during 2008 in both sexes was 565,650. The breakdown was as follows:

Lung and bronchus cancers killed 161,480
Colon and rectal cancers: 49,960
Prostate cancer: 28,660
Cancer of the esophagus: 14,280
Urinary bladder cancer: 14,100
Kidney and renal pelvis cancers: 13,010
Liver and gall bladder cancers: 18,410
Chronic lymphocytic leukemia: 4,390
Acute lymphocytic leukemia: 1,460
Chronic Myelocytic leukemia: 450
Acute Myelocytic leukemia: 8,820
Other forms of leukemia: 6,590
Non-Hodgkin lymphoma: 19,160
Hodgkin lymphoma: 1,350
Oral cavity/Throat cancers: 7,590
Pancreatic cancer: 34,290
Melanoma: 8,420
Other skin cancers: 2,780
Skin cancer (excluding basal and squamous cells): 11,200
Stomach cancer: 10,880
Small intestine cancer: 1,110
Gall bladder cancer: 3,340
Testicular cancer: 380
Multiple myeloma: 10,690
Breast cancer: 40,930
Ovarian cancer: 15,520
Uterus cancer: 7,470

[1] American Cancer Society Facts and Figures, 2008

Cervical cancer: 3,870
Vaginal cancer: 760
Vulva cancer: 870
Penis cancer: 290
Thyroid cancer: 1,590
Brain cancer: 13,070
Eye cancer: 240[1]

Genetics plays a role in the causation of different cancers; ten to fifteen percent of all cancers are hereditary in nature and the rest are attributable to environmental factors.

Breast Cancer:

In 2008, 184,450 individuals were diagnosed with breast cancer in the U.S.; 1,990 of them were men and 182,460 were women. 40,930 women and men died of breast cancer in 2008 in the U.S. 40,480 women and 450 men died of breast cancer in the in 2008.

Risk factors for breast cancer include:

Being a woman;
BRCA1 and BRCA2 inheritance genes;
Family history of breast cancer (Women who have a family history of breast cancer and who tested negative for BRCA1/BRCA2 genes, have four times the risk of developing breast cancer than do women with no family history of breast cancer).
Fibrocystic breast
Early menarche (menstruation)
Late menopause
Childlessness
Oral contraceptives- use of
Having first child after age 30
Overweight/Obesity
Post menopausal hormone
Alcohol abuse

Signs and symptoms of breast cancer Include:

Lump in the breast

---

[1] American Cancer Society Cancer Facts and Figures, 2008

Pain in the breast
Pain in the nipple
Skin retraction in the breast
Lump seen on mammogram/Sonogram
Abnormal calcifications seen on mammogram/ Sonogram
Abnormality seen on digital mammogram
Abnormal MRI of the breast

Disparity in the incidence and death rates in breast cancer - white women vs. black women.

Even though the incidence of breast cancer is significantly higher in white women as compared to black women, the death rate is higher in black women. Many factors play a role in the higher death rates in black women compared to their white counterpart.

1. The socioeconomic status of black women is worse than that of white women.
2. More black women are uninsured as compared to white women.
3. White women visit physicians earlier and more often than do black women.
4. The incidence of obesity is higher among black women as compared to white women.
5. Black women tend to ignore physical symptoms more than do white women.
6. Thirty-five percent of black women have estrogen/progesterone negative breast cancers as compared to the twenty percent rate of estrogen/progesterone breast cancers in white women. As a rule, ER/PR negative breast cancers have a worst prognosis and higher death rates.

During 200-2004, 132,500 white females per 100,000 had breast cancer in the U.S.; 118.3 black females per 100,000 had breast cancer; 89.0 Asian American/ and Pacific Islanders per 100,000 had breast cancer; 69.8 American Indian and Alaska Natives per 100,000 had breast cancer; and 89.3 Hispanic/Latinos per 100,000 had breast cancer.

The incidence of deaths from breast cancer during the period 2000-2004 in the U.S. per 100,000 individuals was as follows: 25.0 white females; 33.8 black females; 12.6 Asian

American and Pacific Islander; 16.1 American Indian and Alaskan Native; and 16.1 Hispanic/Latinos.[1]

Lung Cancer:

In the year 2008, 215,000 individuals in the U.S. were diagnosed with lung cancer 111,690 men and 100,330 women. During this same period, 161,840 men and 71,030 women died of lung cancer.

There exists great disparity in the incidence of lung cancer among blacks compared to whites and other minorities in the U.S.

In 2000-2004, the incidence of lung cancer per 100,000 men was as follows: 110.6 black males, 81.0 white males, 55.1 Asian American / Pacific Islander males, 53.7 American Indian and Alaska Natives males, and 44.7 Hispanic/Latinos males.

Among women, the incidence was as follows: 54.6 white females, 53.7 black females, 27.7 Asian American and Pacific Islander, 36.7 American Indian and Alaska Natives, and 25.2 Hispanic/Latinos.

In addition to the disparity in the incidence of lung cancer, there is also a significant disparity in lung cancer death rates among whites vs. blacks and other minorities. In 2002-2004, the numbers for men were as follows: 95.8 black males, 72.6 white males, 38.3 Asian American and Pacific Islander males, 49.6 American Indian and Alaska Natives males, and 36.0 Hispanic/Latinos males.

For women, the breakdown was as follows: 42.1 white females, 39.8 black females, 18.5 Asian American and Pacific Islander, 32.7 American Indian and Alaska Natives, and 14.6 Hispanic /Latinos.[2]

The most common reason why people develop lung cancer is tobacco smoking which includes exposure to second hand smoke. Exposure to industrial toxins such as coal emission, pesticides, etc, is also a common causal agent.

Some of the symptoms of lung cancer include coughing, shortness of breath, weight loss, etc.

---

[1]   American Cancer Society, Surveillance Research, 2008.
[2]   American Cancer Society Cancer Facts and Figures, 2008

Risk factors for lung cancer include cigarette and cigar smoking, exposure to second-hand smoke, exposure to industrialized fumes such as toxic chemicals, arsenic, and air pollution from coal-burning machines, exposure to asbestos, scars from healed tuberculosis or other inflammatory lung diseases, genetic predisposition to cancer, etc.

The incidence of tobacco smoking is quite high among blacks and other minorities. Because blacks and other minorities have poorer economic, educational, and psychosocial standing in society, and are exposed to working and living conditions that are so much worse than that of whites, that both the incidence and death rates from lung cancer are much higher in minorities than in their white counterparts.

Colorectal Cancer

In the year 2008, 148,810 individuals were diagnosed with colorectal cancer in the U.S. Of this number, 77,250 were men and 71.560 were women. The number of people who died during that year was 49,960 - 24,260 men and 25,700 women.

There exists significant disparity in both the incidence and death rate of colorectal cancer in white men and women vs. black men and women.

In 2004, according to the ACS, the numbers were as follows per 100, 000 individuals: 60.4 white males and 44.0 white females had colorectal cancer; 72.6 black males and 55.0 black females had colorectal cancer; 22.9 white males and 15.9 white females died of colorectal cancer and 32.7 black males and 22.9 black females died of colorectal cancer.

Poverty, lower level of education, poor housing conditions, work related discrimination and the overall racial discrimination and bigotry that blacks have to deal with on a constant basis plays a major role in both the higher incidence of cancer and the higher death rate that is seen in blacks and other minorities in the U.S. Because of poverty, blacks and other minorities are often forced to eat a diet that is rich in fats, salt, spice and simple carbohydrates. This diet causes obesity and obesity increases the risk of developing cancer of the colon. The high salt content of such a diet causes water retention which also causes obesity. The high spice content of this diet causes irritation of the rectal tissue increasing the risk of the development of cancer of the rectum. The high simple carbohydrates content of this type of diet causes the conversion of fat in the liver from simple carbohydrates which is deposited in different tissues in the body also contributing to obesity.

The high fat content of the so called "soul food" and "fast food" diets causes the deposition of large quantities of fat into the colon. When the fat gets into the colon, it requires the work of colonic bacteria to break it down to complete the digestive process.

The adult human colon contains 100 trillion bacteria and 100 billion different species of

bacteria. When these bacteria break down the fat in fatty foods, the release of a series of toxic chemical materials occur. These toxic chemical materials cause the development of an irritating reaction in the wall of the colon. This local irritation causes the development of an inflammatory reaction. The colonic inflammation causes the development of polyps of different types and sizes which can ultimately result in colon cancer.

The incidence of cancer is 37% higher in black males as compared to white males and the incidence of cancer is 17% higher in black females as compared to white females. The 5 year survival rate from cancer is 57% for Blacks as compared to 67% in Whites.

The incidence of liver cancer in Latino men and women is twice that of white men and women. Overall, Latinos have a lower incidence of all other cancers as compared to whites. Asian Americans and Pacific Islanders have a lower incidence of all cancers than whites except for cancer of the liver and stomach. The death rates for cancer of the liver and stomach is higher for Asian Americans and Pacific Islanders. The incidence and death rate from cancer of the kidney in American Indians and Alaska Natives is higher than any other ethnic group.

Cancer of the Pancreas

Cancer of the pancreas usually develops symptom free until it is too late. In 2008, 37,680 individuals were diagnosed with cancer of the pancreas in the U.S. Of this number, 18,770 were men and 18,910 were women.

During this same period, 34,290 individuals died of pancreatic cancer. Of this number, 17,500 were men and 16,790 were women. According to the ACS in 2003, 16.2 black men per 100.000 were diagnosed with cancer of the pancreas and 12.7 white men per 100.000 were diagnosed with cancer of the pancreas. A difference of 3.5 more black men per 100.000 were diagnosed with pancreatic cancer as compared to white men. During this same period, 15.4 black men per 100.000 died of cancer of the pancreas as compared to 12.0 white men per 100,000 who died of cancer of the pancreas.

These statistics shows a clear disparity of 3.4 more black men per 100.000 dying of cancer of the pancreas as compared to white men. 13.7 black women per 100.000 were diagnosed with cancer of the pancreas and 9.8 white women were diagnosed. Therefore, 3.9 more black women were diagnosed with cancer of the pancreas as compared to white women. 12.5 black women per 100.000 died of cancer of the pancreas as compared to 9.0 per 100.000 white women. Thus, 3.4 more black women per 100.000 died of pancreatic cancer as compared to white women.

Risks for cancer of the pancreas include tobacco smoking, chronic pancreatitis, diabetes mellitus, a high fat diet, cirrhosis of the liver, hepatitis B. and obesity.

"Obesity during early adulthood is associated with a greater risk of pancreatic cancer and a younger age of the disease. Obesity at an older age is associated with a lower overall survival in patients with pancreatic cancer"[1]

There are many possible reasons that explain the higher incidence and death rate in pancreatic cancer in blacks as compared to whites.

1. The incidence of tobacco smoking is higher in blacks as compared to whites;
2. The incidence of pancreatitis is quite high among blacks secondary to alcoholism;
3. Gall stone disease (more blacks suffer from gall stone than do whites);

---

[1]   JAMA, Vol. 301.No 24, June 24, 2009.

4. Sickle cell disease and obesity (both conditions are more prevalent in blacks);

5. The incidence of alcohol abuse is much higher in blacks as compared to whites in the U.S.;

6. The incidence of diabetes mellitus is 3-4 times higher in blacks as compared to whites;

7. The diet of most blacks is higher in fat content than of whites;

8. The incidence of cirrhosis is much higher in blacks as compared to whites;

9. The incidence of hepatitis B is much higher in blacks than in whites;

10. More blacks are guilty of Intravenous Drug Abuse than whites causing them to develop hepatitis B more frequently and thereby increasing their incidence of pancreatic cancer.

Cancer of the Mouth, Throat, and Neck

Cancer of the oral cavity and pharynx is most common in tobacco smokers and heavy alcohol users, individuals who have had radiation to the neck area to treat enlarged tonsils, or individuals who have been exposed to industrialized cancer causing materials, and individuals who have been exposed to HPV during oral sex.

In 2008, 35,310 individuals were diagnosed with oral cavity and pharynx cancer. Of this number, 25,310 were men and 10,000 were women. During this same period, 7,590 individuals died of this cancer; 5,210 were men and 2,380 were women.

There exists significant disparity in both the incidence and death from oral cavity and cancer of the pharynx in blacks as compared to whites. In 2003, according to the ACS, the numbers of individuals diagnosed per 100,000 individuals were as follows: 18.0 black males compared to 15.7 white males; that is, 2.3 more black males were diagnosed with oral cavity compared to white males. 6.8 black males per 100,000 died of oral cavity and pharynx compared to 3.8 white males, a difference of 3.0 black men vs. white men dying from the disease. 5.8 Black females per 100,000 were diagnosed with oral cavity and pharynx cancer and 6.1 white females per 100,000 were diagnosed with oral cavity and pharynx cancer; 1.7 black females per 100,000 died of oral cavity and pharynx cancer as compared to 1.5 white females who died from the same disease.

Some signs and symptoms of mouth, throat, and neck cancers include:

1. A sore that bleeds often and easily and fails to heal;
2. Persistent hoarseness that is not accompanied by pain;
3. Change in voice from normal to a persistent deeper voice;
4. Persistent hoarseness that is accompanied with pain;
5. Difficulty in swallowing;
6. Difficulty in chewing, swallowing, and moving the tongue; and
7. Painful swelling around the neck.

Some risk factors for mouth and throat cancer are smoking cigarettes, cigars or a pipe, as well as heavy alcohol consumption and radiation around the neck area during childhood. Oral sex predisposes those who participate in it because of possible HPV infection of the genitalias.

Cancer of the Urinary Bladder:

Cancer of the urinary bladder was diagnosed in 68, 810 individuals in 2008 in the U.S. Signs and symptoms of cancer of the bladder are hematuria (blood in the urine) and frequent urination and burning on urination.

Some risk factors associated with cancer of the bladder are cigarette smoking, chemicals used in rubber, leather, dye used in industries and materials used in hair products, exposure to Cytoxan chemotherapy, history of external radiation to the rectum, decreased fluid intake over an extended period, exposure to schistosoma heamatobium parasite, etc.

Cancer of the urinary bladder is more common in white males and females as compared to black males and females. In 2003, according to the ACS, the numbers per 100,000 individuals were as follows: 19.8 black males were diagnosed with the disease compared with 40.2 white males. That is, a difference of 20.4 more white males had cancer of the urinary bladder during that year. 7.4 black females compared to 10.0 white females were diagnosed with cancer of the urinary bladder. That is, a difference of 2.6 more white females had cancer of the urinary bladder. 5.4 black females per 100.000 died of cancer of the urinary bladder and 7.8 white females per 100.000 died of the disease. That is, a difference of 2.6 more white females died of cancer of the urinary bladder in 2003.

Cancer of the Kidney:

In 2008, the ACS reported that 54,390 individuals in the U.S. were diagnosed with cancer of the kidney and during that same period, 13,010 died of this cancer. During this same year, 33,130 men were diagnosed with cancer of the kidney and 21,260 women were diagnosed with the disease. The number of men who died of the disease was 8,100 and the number of women was 4.910. The breakdown of black and white males was 20.1 black males and 18.0 white males per 100,000 individuals. That is, a difference of 2.1 more black males per 100.000 as compared to white males had cancer of the kidney during 2008.

In 2003, 9.7 black males per 100.000 had cancer of the kidney and 9.0 white males had the disease. That is, a difference of 0.7 more black males per 100.000 had cancer of the kidney as compared to white males in that year. In that year, the death rate was as follows per 100,00 individuals: 6.2 black males and the same proportion, 6.2 white males died of cancer of the kidney. 2.8 black females per 100.000 died of cancer the kidney in 2003 and the same proportion, 2.8 white females per 100.000 died of cancer of the kidney during that same period of time.[1]

Quite often, cancer of the kidney is discovered incidentally. Early stage cancer of the kidney usually presents with no symptoms. Frequently, blood is found during a routine urinalysis which triggers an evaluation resulting in establishing the diagnosis. Kidney cancer may also be diagnosed when a person presents with either microscopic hematuria (Red blood cells seen under the microscope) or gross hematuria blood seen by the naked eyes. Low back pain and flank pain are also common symptoms of the disease.

---

[1] American Cancer Society: *Cancer Facts & Figures for African Americans*, 2007-2008

Testicular Cancer:

In 2008, 8,090 men were be diagnosed with testicular cancer in the U.S. and during that same period 380 men died of this cancer. 1.5 black males per 100.000 had testicular cancer in 2003 in the U.S. and 6.2 white males per 100.000 had testicular cancer. A difference of 4.7 per 100.000 less black men had Testicular cancer as compared to white males. Therefore, testicular cancer is almost 5 times more common in white males than in black males. (Source: Cancer Facts & Figures for Americans 2007-2008)

Early testicular cancer presents with no symptoms. The first sign of testicular cancer is a palpable mass in the testicle. Usually, the mass is felt by the affected person and confirms by an examining physician.

Prostate Cancer:

In 2008, it is reported that 186,320 men were diagnosed with prostate cancer and 28,660 men died of prostate cancer. In 2003, 258.3 black males per 100.000 had prostate cancer in the U.S. and 163.4 white males per 100.000 had prostate cancer during that same period. That is, a difference of 94.9 more black males per 100.000 had prostate cancer than white males in 2003. The death rate was 64.0 black males per 100.000 and 26.2 white males per 100.000 during that same time. That is, a difference of 37.8 more black males per 100.000 died of prostate cancer as compared to white males. A difference of 2.4 more black males died of prostate cancer than did white males in 2003.

The risk factors for prostate cancer include being black-from any country, being black American, being Jamaican black and being a man from any ethnic group, being obese, and having certain genetic factors.

Black American Men and Black Jamaican men have the highest risk of developing prostate cancer than other men in the entire world. They not only develop prostate cancer at an earlier age, (from age 35 onward), but their prostate cancer is more aggressive and harder to treat. The death rate from prostate cancer in this group of men is the highest of all other groups of men in the world.

The symptoms and signs of prostate cancer include difficulty passing urine, urinary frequency, burning on urination, blood in the urine, urinary tract infection, elevated prostatic specific antigen (PSA), an enlarged, hard, and irregular prostate gland on digital rectal examination, a palpable prostate nodule on digital rectal examination - normal PSA is 0-4.0, depending on the person's age.

The PSA can be elevated because of urinary tract infection, benign prostatic hyper trophy (enlarged prostate), and ejaculation. A PSA can be 1.0 and yet the man may be harboring prostate cancer. This is the reason it is very important for men who are at risk for prostate cancer to have an annual digital rectal examination. In this way, the prostate gland can be palpated to rule out a nodule, etc.

Leukemias:

In 2008, 44,270 individuals were diagnosed with new leukemia in the U.S. - 25,180 males and 19,090 females. Of this number, 25,180 males and 19,090 females died of leukemia during that period.

In 2003, 12.9 black males per 100.000 had leukemia and 16.5 white males per 100.000 had leukemia. A difference of 3.6 less black males per 100.000 had leukemia in 2003 as compared to white males. In that same year, 8.0 black females per 100.000 had leukemia and 9.8 white females per100.000 had the disease. That is, a difference of 1.8 less black females had leukemia in 2003 as compared to white females. [1] Overall, leukemia is more common in whites than in blacks.

There are eleven different types of leukemia:

Acute lymphocytic leukemia
Chronic lymphocytic leukemia
Acute myelogenic leukemia
Chronic myelogenic leukemia
Monocystic leukemia
Myelodysplastic syndrome
Acute megakaryocytic leukemia

---

[1] American Cancer Society, *Cancer Facts & Figures for African Americans*, 2007-2008

Acute Promyelocytic Leukemia
T-cell leukemia/lymphoma due to HTLVI and II.
Burkitt's Leukemia
Hairy cell Leukemia

Signs and symptoms of leukemia include weight loss, fatigue, frequent infections, easy bruising, nosebleeds, and hemorrhages.

Some of the risk factors include Down's syndrome, AIDS, exposure to ionizing radiation, chemicals like benzene and other toxic chemicals, and viruses.

Lymphoma:

In 2008, 74,340 individuals in the U.S. were diagnosed with lymphoma. Of these cases, 66,120 were Non-Hodgkin lymphoma and 8,220 were Hodgkin's lymphoma. Of this number, 39,850 males and 34,490 females had lymphoma. In this same year, 35,450 males had Non-Hodgkin lymphoma, 30,670 females had Non-Hodgkin lymphoma, 4, 400 males had Hodgkin's lymphoma and 3,820 females had Hodgkin's lymphoma. In 2008, 20,510 individuals died of lymphoma in the U.S., 10,490 of them were males and 10,020 were females.

In 2003, 2.8 black males per 100.000 had Hodgkin's lymphoma and 3.2 white males per 100.000 had Hodgkin's lymphoma in the U.S. In the same year, 17.6 black males per 100.000 had Non-Hodgkin lymphoma in the U.S. and 23.8 white males had Non-Hodgkin lymphoma. That is, a difference of 6.2 less black males had Non-Hodgkin lymphoma as compared to white males. Also in 2003, 2.0 black females per 100.000 had Hodgkin lymphoma and 2.6 white females had Hodgkin lymphoma. In this same year, 11.7 black females had Non-Hodgkin lymphoma and 16.8 white females had Non-Hodgkin lymphoma. That is, a difference of 5.1 less black females per 100,000 had Non-Hodgkin lymphoma as compared to white females. In addition, in 2003, 0.5 black males per 100.000 died of Hodgkin lymphoma and 0.6 white males per 100.000 died of Hodgkin lymphoma.

In 2003, the numbers for females per 100,000 were as follows: 4.3 black females and 6.5 white females died of Hodgkin lymphoma. That is, a difference of 2.2 less black females died of Hodgkin lymphoma as compared to white females. In 2003, 0.3 black females died of Non-Hodgkin lymphoma and 0.4 white females died of Non-Hodgkin lymphoma. These statistics show that overall more whites have lymphoma than blacks do, and more whites die of lymphoma than blacks do.

Some of the signs and symptoms of lymphoma are enlarged lymph nodes, fever, and weight loss, loss of appetite, night sweats, anemia, and sometimes diarrhea. Sometimes a fever can come and go for several weeks—referred to as "fever of unknown origin" (FUO).

Multiple Myeloma:

Multiple myeloma is a cancer that comes from plasma cells, cells that produce antibodies in the body to help fight infections. In 2008, 19,920 individuals were diagnosed with multiple myeloma in the U.S. and 10,690 of them died of multiple myeloma during that period. Of this number, 11,190 were males and 8,730 were females. In 2008, 5,640 males and 5,050 women died of multiple myeloma.

In 2003, it was reported that 13.7 black males per 100.000 had multiple myeloma in the U.S. and 6.5 white males - a difference of 7.2 more black males per 100.000 had multiple myeloma as compared to white males in 2003.The numbers for women per 100,000 in 2003 were as follows: 9.1 black females and 4.1 white females - a difference of 5.0 more black females per 100.000 had multiple myeloma as compared to white females in 2003.

In 2003, the death rates per 100,000 individuals from multiple myeloma were reported as follows: 8.5 black males and 4.4 white males - a difference of 4.1 more black males per 100.000 died of multiple myeloma in the U.S. in 2003. In 2003, the death rates in females from this disease were as follows: 6.3 black females and 2.9 white females per 100.000 died of multiple myeloma. That is, a difference of 3.4 more black females died of multiple myeloma in 2003 in the U.S. as compared to white females.[1]

---

[1] American Cancer Society, *Cancer Facts & Figures for African Americans*, 2007-2008.

There exist no clear risk factors for multiple myeloma, although exposure to agricultural chemicals, radium, benzene, and radioisotopes have all been mentioned as associated with a higher incidence of multiple myeloma.

Overall, the incidence of multiple myeloma is several times higher in blacks compared to whites. More blacks also die of multiple myeloma than do whites.

One can make the case that since blacks work more frequently in jobs that expose them to more toxic materials than do whites, this may explain their higher rate of developing multiple myeloma and thus their higher death rate.

Some of the symptoms and signs of multiple myeloma are bone pain, weakness, anemia, recurrent infections such as pneumonia, osteoporosis, kidney failure, and high serum calcium and its associated problems such as seizures, etc. It is said that Interleukin 6 (a growth factor) is the causative protein that supports the growth of myeloma cells.

Basal Cell Carcinoma of the Skin:

The most common cancers of the skin are basal cell and squamous cell carcinomas (skin cancer) affecting about one million people yearly in the United States. This form of cancer appears most frequently in people who have fair skin and people who are exposed more frequently to the sun.

Signs and symptoms of these types cancer include skin sores that bleed frequently and will not heal. A Dermatologist should be seen to evaluate and biopsy these lesions.

In 2008, 67,720 people in the U.S. were diagnosed with basal cell carcinoma and 11,200 people died of the disease during that same year. In 2008 also, 38,150 males and 29,570 females had basal cell carcinoma. In that year, 11,200 males and 7,360 females died of basal cell carcinoma.

Melanoma:

The most serious and malignant form of skin cancer is melanoma and in 2008, there were 62,480 cases of melanoma diagnosed in the U.S. and 8,420. Of these cases, there were 8,420 fatalities.

Signs and symptoms of melanoma are a mole or a sore on the skin that bleeds and will not heal. A Dermatologist should be seen immediately to evaluate and biopsy these lesions.

In 2008, it was reported that 34,950 males and 27,530 females had melanoma in the U.S. During that same year, 5,400 males and 3,020 females died of the disease.

Melanoma is ten times more common in whites than in people of color.

In 2003, it was reported that 1.1 black males per 100.000 had melanoma in the U.S. and 26.5 white males per 100.000 had melanoma during that same time. A difference of 25.4 more white males per 100.000 had melanoma than black males.

In 2003, 0.9 black females per 100.000 had melanoma in the U.S. and 17.3 white females per 100.000 had melanoma during that same period. A difference of 16.4 more white females had melanoma as compared to black females.

In 2003, 0.5 black males per 100.000 died of melanoma in the U.S. and 4.3 white males per 100.000 died of melanoma during that same time. That is, a difference of 3.8 more white males per 100.000 died of melanoma as compared to black males.

In 2003, 0.4 black females in the U.S. and 2.0 white females died of melanoma during that same time. A difference of 1.6 more white females died of melanoma as compared to black females.[1]

Thyroid Cancer:

In 2008, 37,340 individuals were diagnosed with cancer of the thyroid gland in the U.S. and 1.590 of them died of the disease. In 2008, it was reported that 930 males and 28,410 females had thyroid cancer in the U.S. In 2008. 680 males and 910 females died of cancer of the thyroid in the U.S.

In 2003. 2.4 black males and 4.5 white males per 100,000 had cancer of the thyroid. That is, in that year, a difference of 2.1 more white males had cancer of the thyroid as compared to black males in the U.S. During that same time, 7.1 black females per 100,000 and 12.7 white females per 100,000 had cancer of the

[1] American Cancer Society, *Cancer Facts & Figures for African Americans*, 2007-2008

thyroid. A difference of 5.6 more white females per 100.000 had cancer of the thyroid as compared to black females.

In 2003, 0.4 black males and 0.5 white males per 100,000 died of cancer of the thyroid in 2003 in the U.S. A difference 0.1 more white males per 100,000 died of cancer of thyroid as compared to black males. During that same year, 0.5 white females and 0.5 black females per 100,000 died of cancer of the thyroid in the U.S. Overall, cancer of the thyroid is more common in whites as compared to blacks.[1]

Cancer of the Esophagus:

In 2008, 16,470 individuals were diagnosed with cancer of the esophagus in the U.S. Of this number, 12,970 were men and 3,500 were women. Fatalities from this disease during this period were 14,230 individuals, 11,250 men, and 3,030 women.

In 2003, 10.8 black males and1.9 white males per 100,000 had cancer of the esophagus in the U.S. A difference of 3.0 more black males per 100,000 had cancer of the esophagus as compared to white males. In the female population, 3.3 black females and 1.9 white females per 100,000 had cancer of the esophagus during that same time. A difference of 1.4 more black females per 100,000 had cancer of the esophagus as compared to white females.

In 2003, 10.5 black males and 7.7 white males per 100,000 died of cancer of the esophagus. A difference of 2.8 more black males died of cancer of the esophagus as compared to white males. Among women, 3.0 black females per and 1.7 white females per 100,000 died of cancer of the esophagus in 2003 in the U.S. A difference of 1.3 more black females per 100,000 died of cancer of the esophagus as compared white females.

More blacks have cancer of the esophagus than do whites most probably because more blacks per capita smoke tobacco and drink alcohol than whites do.

In addition, the diet of most blacks is fattier and spicier than the diet of most whites. Alcohol is very irritating to the lining of the esophagus. The fattier and spicier the diet, the more irritated his or

---

[1] American Cancer Society, *Cancer Facts and Figures for African Americans*, 2007-2008

her esophagus becomes. The irritation of the esophageal lining can lead to dysplasia of tissues in the esophageal lining and over time, the dysplasia can lead to DNA mutation causing cancer to develop.

Therefore the disparity that exists in the life style in blacks vs. whites plays a major role in the higher incidence and deaths that occur in blacks as compared to whites.

Cancer of the Stomach:

In 2008, 21,500 individuals were diagnosed with cancer of the stomach in the U.S. Of this number, 13,190 were men and 8,310 were women. During that same time, 10,880 individuals died of this cancer, 6,450 men and 4,430 women.

In 2003, 17.7 black males and 10.2 white males per 100,000 had cancer of the stomach in the U.S. A difference of 7.5 more black males per 100,000 had cancer of the stomach as compared to white males.

In 2003, 9.3 black females and 4.7 white females per 100,000 had cancer of the stomach. A difference of 4.6 more black females had cancer of the stomach as compared to white females. In the male population, 12.1 black males and 5.3 white males per 100,000 died of cancer of the stomach. A difference of 6.8 more black males per 100,000 died of cancer of the stomach as compared to white males.

In 2003, 17.7 black females and 2.7 white females per 100,000 died of cancer of the stomach. A difference of 3.3 more black females per 100,000 died of cancer of the stomach as compared to white females.

There exists a clear disparity both in the incidence and death rate of cancer of the stomach in blacks compared to whites. One of the reasons for this lies in the diet of most blacks which consists more fried, greasy and spicy foods. No doubt that the state of poverty that afflicts most black people in the U.S. plays a major role in both the higher incidence and higher death rate of cancer of the stomach in blacks as compared to whites.

Cancer of the Liver:

In 2008, 21,370 individuals were diagnosed with cancer of the liver in the U.S. Of that number, 15,190 were men and 6,180 were women. Of that number, there were 18,410 deaths. 12,570 were men and 5,840 were women.

In 2003, 12.1 black males and 7.8 white males per 100,000 had cancer of the liver. A difference of 4.3 more black males per 100,000 had cancer of the liver in 2003 in the U.S. as compared to white males.

In 2003, 3.5 black females and 2.8 white females per 100,000 had cancer of the liver. A difference of 0.7 more black females had cancer of the liver as compared to white females in 2003 in the U.S.

In the male population in 2003, 9.8 black males and 6.4 white males per 100,000 died of cancer of the liver in the U.S. A difference of 3.4 for more black males per 100,000 died of cancer of the liver in 2003 as compared to white males.

In 2003, 1.8 black females and 2.8 white females died of cancer of the liver in the U.S. A difference of 1.0 more black female per 100,000 died of cancer of the liver as compared to white females during that same time.[1]

Several possible reasons exist to explain this high degree of disparity in blacks'

Vs. whites' incidence of liver cancer and deaths from liver cancer including:

1. The rate of cirrhosis of the liver is higher in blacks as compared to whites. Cirrhosis of the liver can lead to liver cancer.

2. More blacks get infected with hepatitis B and C with intra-venous illicit drug use as compared to whites. Both hepatitis B and C can cause cancer of the liver. In addition, more blacks get exposed to cancer causing toxic materials at the work place than do whites. Because of racial discrimination, blacks typically get the type of jobs that are more likely to expose them to toxic materials at the workplace.

---

[1] American Cancer Society, *Cancer Facts and Figures for African Americans,* 2007-2008

CHAPTER 10

DISPARITY IN KIDNEY DISEASES IN THE U.S.

The incidence of end-stage renal disease (ERSD) is higher in blacks than that of all other racial groups in the U.S. There are roughly 370,000 individuals suffering from kidney failure in the U.S.

The risk factors for ESRD are:

1. Hypertension
2. Diabetes mellitus
3. Acute renal failure because of hypotension resulting from acute myocardial infarction as seen in individuals who are hypertensive and diabetic or who have had hypotensive episodes during gastrointestinal bleeding, sepsis, etc.
4. Chronic pyelonephritis
5. Sickle cell anemia
6. Polycystic kidney disease
7. Glomerulonephritis
8. Thrombotic Thrombocytopenic Purpura (TTP)
9. Hemolytic Uremic Syndrome
10. Kidney Stones
11. Lupus Nephritis
12. Rhabdomyolysis
13. Acute tubular necrosis (ATN) etc.

Diabetes is the number one cause of end-stage renal disease in the U.S.

Hypertension is a number two cause of end-stage renal disease in the U.S.

Hypertension is one of the leading diseases in the U.S. About 73 million individuals in the U.S. and about 1 billion people worldwide have hypertension. The incidence of end-stage renal disease (ESRD) leading to renal failure is higher in minorities than in whites.

Blacks are 4.7 times more likely than whites to develop kidney failure.

Blacks represent 47% of individuals on dialysis in the U.S. even though they represent only 13.8% of the U.S. population.

The incidence of kidney failure is higher in minorities than in whites. In the year 2000, Blacks had a rate of kidney failure of 777 cases per million. Native- American had a kidney failure rate of 501 cases per million. Asian-American had a kidney failure rate of 281 cases per million. Hispanics/Latinos had a kidney failure rate of 276 cases per million. Whites had a kidney failure rate of 269 cases per million.

The disparity that exists in the occurrence of ESRD in minorities compared to whites in the U.S. is appalling, shameful, and simply unacceptable and must be corrected.Blacks also usually develop ESRD at an earlier age and the kidney failure in blacks is more aggressive.

Twelve percent of adults in the U.S. have chronic renal disease of one degree or another. If chronic renal failure is recognized early and treated properly, it can prevent the development of ESRD.

Chronic renal disease has 5 stages, 1-5.
Stage 1.is GFR 90 ml/minute/1.73m2
Stage 2. is GFR 60-89ml/minute/1.73m2
Stage 3. is GFR 30-59ml/minute/1.73m2
Stage 4 is GFR 15-29ml/minute/1.73m2
Stage 5. is GFR less than 15ml/minute/1.73m2

The most effective treatment for renal failure is a low-salt and low-protein diet. When the renal function deteriorates to the point that the BUN and the creatinine are excessively high, along with high serum potassium, high phosphatase, low calcium and a very low creatinine clearance combined with evidence of uremia, dialysis becomes necessary.

There are two types of dialysis in routine use to treat ESRD: Peritoneal dialysis and Hemodialysis.

Different clinical situations, along with the patient's preference, will help to determine which type of dialysis will be used to treat the individual patient with end-stage renal failure.

Kidney transplant is a significant treatment for suitable patients. Even here, there is a disparity. That is, although blacks and other minorities have a higher incidence of kidney failure, whites receive a higher percentage of kidney transplants than do blacks and other minorities.

Racial discrimination and poverty are key factors in determining who gets kidney transplants in the U.S. In most hospitals, it is usually a Caucasian physician in training who decides who will receive an organ transplants, allowing racial bigotry to create this awful disparity.

CHAPTER 11

DISPARITY IN DISEASES OF THE STOMACH AND INTESTINES IN THE U.S.

DISEASES OF THE GASTROINTESTINAL TRACT are among the most common diseases in the U.S. The most frequent GI symptoms precipitating a visit to the physician include:

1. Heartburn
2. Bitter taste in the mouth
3. Indigestion
4. Bloating
5. Gaseousness
6. Increased flatulence
7. Nausea
8. Vomiting
9. Loss of appetite
10. Easy filling of the stomach when eating (dysphagia)
11. Pain on swallowing food
12. Pain in the stomach area
13. Pain in the abdomen
14. Recurrent diarrhea
15. Rectal bleeding
16. Pain on defecation
17. Hemorrhoids
18. Constipation
19. Diarrhea, etc.

The underlying reasons for these symptoms include:

1. *Hiatal hernia*
2. *Reflux esophagitis*
3. Slow motility of the esophagus
4. Esophagitis due to fungal infection of the esophagus
5. Cancer of the esophagus and other types of cancer
6. Gastroesophageal reflux disease (GERD)
7. Peptic ulcer
8. Helicobacter Pylori gastritis
9. Ulcerative Colitis
10. Cohn's disease
11. Diverticulitis
12. Small bowel obstruction
13. Large bowel obstruction
14. Cancer of the stomach
15. Cancer of the large bowel
16. Parasitic infestation of the large bowel
17. Spastic colon or irritable bowel disease
18. Viral gastroenteritis
19. Infectious gastroenteritis, etc.

Symptoms associated with diseases of the stomach are more common in people of color as opposed to whites and the reasons are many. To start with, the foods that members of minority groups often like to eat have too much fat, salt, carbohydrates, and spices.

In addition, the day-to-day stress that poor minorities have to deal with makes it is easy to see why the incidence of stomach ailments of all sorts is so much higher in minorities than in the people who have a better diet and do not have to deal with the daily and constant stresses of racial discrimination and bigotry. (This has been referred to in part by some one as the **"STATUS SYNDROME"**). [1]

About sixty percent of people in the United States are obese/overweight and obesity is very highly connected with gallstones and gall bladder diseases. Further, many Blacks,

[1] See, JAMA, March 15, 2006-Vol 295, No 11.

Hispanics, Middle Eastern /Mediterraneans and Asians suffer with hemolytic diseases such as sickle cell anemia and sickle thalassemia and thalassemias that predispose them to gall bladder stone disease. People who suffer from these hemolytic anemia produce a substance called bilirubin in excess, which is a pigment that comes from the breakdown products of the hemolyzed red cells. The bilirubin pigment forms bilirubin stones in the gall bladder. Most of the gallstones seen in obese people however are cholesterol stones.

Symptoms of gall bladder stones, which include nausea, vomiting, right-sided abdominal pain—which can at times be referred to the left side of the upper abdomen— can easily be confused with diseases of the stomach such as peptic ulcer or hiatal hernia with reflux.

Symptoms of gall bladder disease can also be confused with diseases of the pancreas. Both acute and chronic pancreatitis can have symptoms that are similar to gall bladder disease. Frequently, a person will present to the doctor with jaundice and no pain. Then, the question becomes is it due to gallstones occluding the common bile duct, resulting in backing up of bile into the bloodstream, or is it due to a tumor (usually cancer) at the head of the pancreas pressing on the common bile duct, causing the jaundice to occur - a condition called painless jaundice?

Another common cause of abdominal pain is disease of the pancreas. The most common disease of the pancreas is acute pancreatitis. The reason that acute pancreatitis is so common in people of color is because 50% of African-Americans, 46% of Hispanics and 35% of whites are obese, and obesity increases the incidence of gallstones. Gallstone disease is the second most common cause of pancreatitis. Gall bladder and gallstones diseases also have quite a high incidence in American Indians and Eskimos.

The small bowel and large bowel can cause pain and many other symptoms to develop. Among them are ulcerative colitis and Crohn's disease. Both of these conditions are reasonably common in people of color. Some of the first signs are rectal bleeding, cramps, diarrhea, pain, iron deficiency anemia, etc.

No one knows what causes Crohn's disease and ulcerative colitis. To diagnose inflammatory bowel diseases, both barium studies and colonoscopic examinations are used. To diagnose inflammatory bowel of the small bowel, barium, capsule study or scoping are needed.

Diagnosis of inflammatory bowel disease of the large intestine can be made both by barium studies and by colonoscopic examinations. In inflammatory bowel disease, the inner surface of the bowel is swollen and inflamed and bleeds easily. The cause or causes of these changes are not known in spite of many years of research.

In addition to abdominal pain, diarrhea, and rectal bleeding, there is an increased incidence of colorectal cancer in many people who are afflicted with inflammatory bowel diseases. Among the symptoms of inflammatory bowel diseases are:

1. Diarrhea
2. Constipation
3. Diarrhea alternating with constipation
4. Abdominal cramps
5. Flatulence
6. Abdominal pain
7. Rectal bleeding
8. Ulcerative colitis
9. Crohn's disease of the colon
10. Diverticulosis
11. Diverticulitis
12. Bacterial overgrowth
13. Lactose intolerance
14. Acute infectious gastroenteritis
15. Parasitism
16. Ischemic colitis
17. Intestinal obstruction
18. Colon cancer
19. Rectal cancer
20. Rectal fissures
21. Hemorrhoids
22. Inguinal hernias
23. Familial polyposis.

Constipation can be caused by:

1. Stress
2. Poor eating habits
3. Hypothyroidism
4. Taking laxatives too often, resulting in a condition called cathartic colon, that is to say, the colon has lost its ability to contract properly because the individual is compulsively abusing laxatives to bring about daily bowel movements. It is not necessary to have a bowel movement every day. A bowel movement every other day is perfectly fine.
5. Constipation due to medications. Good examples of medications that can cause constipation are the calcium channel blockers. The very reason that these medications work to bring down blood pressure is that they relax the smooth muscles within the vessels. The intestines have smooth muscles in them and once these smooth muscles are relaxed, the bowel is likely to lose its contractile force, resulting in constipation. Fortunately, these very important medications do not cause this problem in everyone who takes them.
6. Irritable bowel syndrome, a condition associated with spasm of the bowel is frequently associated with abdominal cramps and constipation.
7. The most feared condition sometimes seen in people who are constipated is cancer of the large bowel. Cancer of the large bowel can cause constipation by mechanically preventing stool from passing through the area where the cancer is, resulting in pencil-sized stools and straining during defecation.

Abdominal pain can be due to many things, including acute appendicitis, acute peritonitis due to conditions such as intra-abdominal abscess of different types, ischemic colitis, ulcerative colitis, cancer, peptic ulcer, perforated peptic ulcer, gall bladder disease, acute and chronic pancreatic diverticulitis, and kidney stones. Other common causes of abdominal pain are acute gastroenteritis due to viral, bacterial, fungal, and parasitic and protozoal infections.

People in Africa and other third world countries suffer only a fraction of the colon cancer than Americans. Dietary habits have a lot to do with minority health problems including, as already stated but bears repeating, too many foods such as red meat, bacon, eggs, sausages all of which result in too much fat and too little grains, fruits and vegetables. The "soul food" tastes good but it is not healthy food. The inability to afford foods that are healthy is quite real and remains a serious problem, which, in today's economic climate is getting worse instead of better. As the economic situation of minority people in this country worsens, the health of minorities will also continue to worsen. The incidence of colon cancer is going to get worse as the diet of minority people gets poorer in the United States. The poorer diet reflects directly on the poor economic status of minorities as compared to whites.

Colorectal cancers are common in all groups in the United States. These cancers are more common in minorities as compared to whites. This may be attributable to many reasons, and prominent among these reasons is the fat-rich diet of many minorities and the fact that the rate of obesity/over weight is higher among minorities as compared to whites. In addition, the fact that minorities, as a rule, go less frequently to physicians to be examined plays a major role. By the time a person develops symptoms such as abdominal pain, nausea, vomiting, diarrhea, rectal bleeding because of colorectal cancer, often the cancer is already in an advanced stage. If the person is lucky, the rectal bleeding might be due to a precancerous polyp or some other nonmalignant lesion.

It is therefore always necessary to pay close attention to the complaints of rectal bleeding or worsening of pre-existing hemorrhoids by undertaking a gastrointestinal evaluation of the lower gastrointestinal tract by a skilled gastroenterologist to be certain that no underlying cancerous mass is causing the obstruction.

Inguinal hernia can be associated with colorectal cancer. An obstructing mass within the large bowel inevitably causes the person harboring the mass to generate a great deal of pressure in the muscle of the lower abdomen. That set of interactions can result in tearing of intra-abdominal muscle causing the development of inguinal hernia. It is therefore very important to investigate a person who is in the cancer age group, age 45 and older, who spontaneously develops an inguinal hernia. Any person who fits that profile ought to have a lower GI evaluation with either a barium enema or a colonoscopy before he undergoes an inguinal hernia repair.

Colonoscopy is recommended to start at age 40 in Blacks and Hispanics because of the higher rates of colon cancer in these two minority groups. In addition, colon cancer is being seen at earlier ages in these two groups.

It is important to keep in mind that cancer of the right colon is missed six percent of the time during a colonoscopy, increasing the rate of colon cancer deaths.[1] Overall, colonoscopy misses about two percent of colon cancer. While the rate of deaths from colon cancer is decreasing among whites, it is rising among minorities, exposing another glaring example of the health disparities that minorities endure in the U.S.

There exists the common perception that blacks abuse alcohol more than whites.

In fact, that is not true. White men and women abuse alcohol more than blacks.

Among all the minorities, Native American Indians/Alaskan Natives tend to have a higher rate of alcohol abuse than the other minority groups. This is so because the rate of poverty among Native American Indians/Alaskan Natives is probably the highest in the U.S. Poverty leads to despair and depression which then causes a resort to alcohol to ease the pain of the living conditions. Unfortunately, the result is a high incidence of alcoholism, pancreatitis, liver disease, depression, spousal abuse, a high rate of homicide and suicide.

---

[1] Annals of Internal Medicine, Vol. 150, Number 1, January 6, 2009.

CHAPTER 12

DISPARITY IN ANEMIA IN THE U.S.

Anemia is common medical problem. The most common underlying reasons for anemia are:

1. Bleeding
2. Abnormal hemoglobin (hemolysis)
3. Autoimmune hemolytic anemia
4. AIDS
5. Starvation
6. Chronic infection
7. Chronic diseases
8. Cancer
9. Chemotherapy
10. B12 deficiency
11. Folic acid deficiency
12. Iron deficiency
13. Chronic kidney failure
14. Arthritis
15. Collagen vascular diseases
16. Alcoholism
17. Malaria
18. Parasitic infestation
19. Iron overload diseases etc;

Bleeding can cause anemia as a result of:

1. Trauma
2. Gastrointestinal bleeding

3. Bleeding from the kidneys
4. Bleeding from the urinary bladder
5. Inflammatory bowel disease
6. Bone marrow failure
7. Menstrual blood loss
8. Thrombopathy
9. Thrombasthenia
10. Coagulopathy
12. Blood loss from multiple child births
13. Abnormal vaginal bleeding, etc.

Still another common anemia seen in many poor people is nutritional deficiency anemia. Many poor people are forced by virtue of their impoverished condition not to eat nutritious foods because these foods are, for the most part, expensive.

Moreover, poor nutrition often begins at an early age, as many poor young people go to school with no breakfast and eat only the food that is provided as part of the school lunch program. The quality and quantity of these foods vary from community to community and in many instances these foods are prototypes of fast foods, which are neither healthy nor nutritious.

Frequently, a combination of these different anemias coexists in minorities making their anemic state much more difficult and complex to diagnose and treat.

According to recent reports, 50 million American citizens live below the poverty line and 24.1% blacks in the U.S. live below the poverty line. It has also been reported that some 14 million children in the U.S. go to bed at night hungry and altogether 38 million individuals suffer from hunger in the U.S. In addition, there are more than several hundred thousands homeless individuals in the U.S. who by virtue of their circumstances are malnourished, and no doubt suffer from anemia of malnutrition.

Blacks and other minorities in the U.S. and elsewhere in the world have a higher incidence of anemia than do their white counterparts. This is so because of many of the reasons mentioned above.

People who migrated to the United States from other parts of the world such as the Americas, the Caribbean, Africa and other tropical countries and in some instances, those who migrated from the rural South of the United States to the north, have the propensity to be infested with parasites. The incidence of parasitic infestation can lead to iron deficiency anemia because of blood loss due to worms sucking blood from the intestines.

Several billion people in the under developed world are infested with worms of one type or another. A large percentage of this population survives on a meager diet to begin with and at the same time, loses a significant percentage of their nutrient intake to parasites which afford them no symbiosis in return. That is clearly an unfair deal. Since it is mostly people of color who populate the underdeveloped world, they have the highest rate of iron deficiency.

Many people use alcohol significantly and alcohol abuse is associated with many conditions that can cause iron deficiency anemia. For instance, alcohol abuse frequently causes gastritis. The blood that is lost because of gastritis can lead to iron deficiency anemia. Alcohol abuse frequently causes esophageal varices (superficial vessels which occur on top of the tissue) with recurrent bleeding, and that too can cause iron deficiency anemia. Alcohol abuse frequently causes damage to the liver resulting in alcoholic liver disease (cirrhosis of the liver) which is one of the causes of esophageal varices. The incidence of colon cancer is quite high in the U.S. and one of the most common signs of colon cancer is iron deficiency anemia.

Black women suffer much more from anemia than do white women. Ninety percent of black women 50 years or older have uterine fibroid and seventy percent of white women age 50 or older have uterine fibroid. Uterine fibroid is associated with heavy menstrual blood loss with secondary iron loss, resulting in chronic iron deficiency anemia.

In addition, black women tend to have more children than do white women and multiple child birth causes iron deficiency anemia.

People of color in general also have a very high rate of being born with abnormal hemoglobin which can cause suffering from life long anemia.

Therefore, it can safely be said that people of color suffer disproportionately from anemia as compared to whites.

This disparity has major negative health significance to the well being of the minority community as a whole in the United States because people who are anemic are chronically sick and cannot function properly to earn a living to provide for themselves and their families.

Sickle Cell Anemia:

A large proportion of Blacks and Hispanics suffer from a sickle cell anemia in the United States. World wide, there are 300 million people who carry the gene that causes sickle cell anemia.

Historically, the sickle cell gene can be traced to three main areas of Africa. The most prevalent sickle cell gene came from Benin near Nigeria in Central Africa. Another gene came from Senegal on the West Coast of Africa. The third gene came from the Bantu-speaking area of Central Africa.

The same three genes can be found within North American Blacks and in the Caribbean. The African slaves who were brought here against their will to work the fields and to do forced labor brought these sickle cell genes to the North American continent during the slave trade. Over the close to 500 years since slavery started, the sickle cell gene has had ample time to penetrate the North American Black race causing much devastation and leaving a lot of pain, suffering, despair and deaths in its wake. As early as 1670, there is evidence that clinical sickle cell disease existed in a Ghanaian family. Sickle cell disease does not only affect Blacks but also some Indians, Italians, and Arabs. In addition, the same percentage of Hispanics throughout the Americas is affected by the sickle cell trait as are Blacks which is 8.5%.

Sickle cell anemia was first diagnosed in the U.S. in 1910 by Dr James Herrick, a cardiologist in Chicago, upon examining a dental student from the island of Grenada by the name of Walter Clement Noel.

There are 2,500,000 individuals in the U.S. who carry the sickle cell trait (SA) and 90,000 individuals carry the full blown gene (SS).

There are four different types of sickle cell disease syndromes:

1. Sickle cell anemia (SS) - the full blown disease;
2. Sickle cell C disease (SC);
3. Sickle cell Thalassemia (S Thal); and
4. Sickle cell trait (SA).

Each of these has its own clinical profile, manifestations, and complications. All people who carry the sickle cell gene suffer significant medical complications from it to one degree or another.

Sickle cell disease is a preventable disease if individuals who carry the gene for this deadly disease would learn the pros and cons of how the disease is inherited. If a person who is not carrying the sickle cell trait gene marries with a person who is carrying the gene for sickle cell trait and they decide to have children, 50% of the children will be born without the sickle cell gene and 50% will carry it. If both individuals are carrying the sickle cell trait, 25% of their children will be born normal, 25% will be born with the full-blown sickle cell disease, and 50% will be born carrying the sickle cell trait. If one of them has the full-blown sickle cell disease and the other is normal, 100% of the children will be born carrying the sickle cell trait. If one of these two individuals is carrying the sickle cell trait (AS) and one has the full-blown sickle cell disease (SS), 50% of the children will be born with the sickle cell trait (AS) and 50% will be born with the full-blown sickle cell disease (SS). If both of these individuals have full-blown sickle cell disease (SS), 100% of their children will be born with full-blown sickle cell disease (SS).

Once a person is carrying either the sickle cell trait (AS) or sickle cell disease (SS), many factors interplay to make that person sick and suffer from sickle cell disease.

Anemia is a serious medical condition and is one of the most common diseases. In some cases, the disease may be prevented with certain precautionary measures.

The disparity that exists in the incidence of anemia between Blacks and other minority as compared to Whites is glaring. The reasons for these glaring disparities include the fact that Blacks are generally poorer than whites. Poverty is highly associated with poor nutrition. Poor nutrition often causes anemia. Blacks living in the developing world have a higher likelihood of being infested with parasites and parasitic infestation causes iron deficiency anemia.

Hereditary hemoglobin diseases are more common in blacks and other ethnic minorities as compared to whites. These abnormal hemoglobin diseases cause anemia.

The rate of colon cancer is higher in blacks and other minorities as compared to whites. Colon cancer causes iron deficiency anemia.

Once it is discovered, a hematologist who is an expert in blood diseases should be consulted to evaluate the anemia properly. Appropriate treatments must then be given to prevent loss of life from anemia and to prevent all the symptoms and complications that are associated with this often deadly disease.

CHAPTER 13

DISPARITY IN UROLOGICAL DISEASES IN THE U.S.

In 2008, 186,320 men in the U.S. were diagnosed with prostate cancer. During that same time, 28,660 men died of prostate cancer.

"Nationally, African-American men are diagnosed with prostate cancer up to 70% more frequently than are white men "[1] The death rates from prostate cancer are 2.4 times higher in African –American and Jamaican men than that of white and Asian men. Other black men living in North America also have higher rates of prostate cancer than whites and Asian men. Similarly, the death rates for these other North-American black men are higher than in both white and Asian men. Other men such as Latinos living in the U.S. and other parts of the North- American continent also have incidence rates for prostate cancer that are higher than white and Asian men.

What are the different factors that predispose men to prostate cancer?

1. <u>Heredity</u> – that is, if a person's father, brother, or uncle has prostate cancer or his mother or aunt has breast, colon or ovarian cancer, the person is likely to also develop a similar type of cancer. There is a 10% genetic crossover between prostate cancer, colon cancer, and ovarian cancer. Certain mothers are capable of transferring the gene for prostate cancer to their sons. Once there is a cluster of cancer in

---

[1] American Cancer Society, 2002

the immediate family, any member of that immediate family has a higher likelihood of developing cancer of one kind or another, more so than in the general population.

2. Recently, several genes have been discovered on chromosome 8 and other loci in the DNA of men who were studied that are said to predispose them to prostate cancer. It would appear that in African-American men, some of these genes may be associated with a more aggressive form of prostate cancer.

3 Obesity – Obesity is associated with the development of prostate cancer because obese men have too much fat in their bodies and are therefore able to use the cholesterol ring associated with fat to overproduce the male hormone-Androgen. The more male hormone a man produces, the more he is able to stimulate the prostate gland, and the more the prostate gland is stimulated by the male hormone, the higher the incidence of developing prostate cancer.

4. Fat-rich diets – Eating too much fat leads to the production of too much male hormone-Androgen - which in turn results in over-stimulation of the prostate gland, resulting in a higher incidence of prostate cancer.

It seems that obese men have prostate cancer with a much lower PSA than do no-obese men because of the dilution factor. It is believed that the fact that the diet of African-American men is too rich in fat explains why the incidence of prostate cancer is the highest in African-American men than all other black men in the world. This author was the first person to have made that observation in 1992 in the author's first book.[1] The medical community has now recognized this as a scientific fact.

Black men from Jamaica, West Indies have the second highest rates of prostate cancer in the world next to African American men. This is most likely due to the fact that the diet of Jamaican men is very high in fat. The rice and beans that many Jamaicans cook is extremely rich in polysaturated fat from the coconut oil with which it is prepared. They cook it this way so that

[1] Alcena,Valiere, M.D., "*The Status of Health of Blacks in the United States – a Prescription for Improvement*", Kendall Hunt Publishing Company (1990).

this rice dish has the distinct taste of the coconut oil. In other parts of the Caribbean, rice and beans is cooked with cooking oil, lard, and pieces of meat - either pork or beef. Another food staple of many Jamaicans is goat meat, commonly called "curried goat". Goat meat is very delicious but is also extremely fatty.

The undeniable fact is that the diet of many African Americans, so-called "soul food" as well as large quantities of fast foods – and the diet of many blacks from the Caribbean forms a significant part of the culture and so is not so easy to eliminate.

It is also true that the fast foods eaten in large quantities by the poor and many minorities are eaten mainly because of their poorer financial circumstances. Some of these fast foods include foods such as cheeseburgers, hamburgers, hot dogs, pizzas, sausages, pancakes often served with syrup and covered with butter, waffles with syrup, fried chicken, and ham, grits covered with butter, egg, and bacon. In addition, there is a high consumption of red meat and pork, foods that are also not healthy, especially when eaten too frequently or in too large quantities.

Black men who live in other parts of the world, such as Africa, have rates of prostate cancer that is many times lower than that of black men who live in North America. Some of the factors which contribute to the much lower rate of prostate cancer in other parts of the world include a diet that is lower in fat and richer in fruits and vegetables. In addition, a diet more rich in poultry, fish and grains, and, a diet more rich in foods such as bananas, potatoes, yams, and plantains. Foods which contain high complex carbohydrates that is not able to be broken down to simple sugar by the body to lead to obesity and also satisfy hunger and provide prolonged energy are better. A lifestyle that includes more physical exercise, including walking and engaging in more manual labor leads to decrease obesity and low incidence of prostate cancer. Finally, the lower incidence of obesity in the developing world also contributes to decrease rate of prostate cancer

Prostate cancer usually appears in black men at age 40. However, there are a few cases known to have occurred as early as age 35. Prostate cancer usually appears in white males at age 50; however, there are a few cases known to have occurred as early as age 45.

Black American men, in particular, as well as other black men must be urged to modify their diet by removing the excessive amount of red meat, pork, fried foods, and fat rich foods and replacing them with non-shellfish, poultry, fruits, vegetables, beans, olive cooking oil and low simple carbohydrate foods, grain, corn meal and other low fat foods. In general, exercise along with a good diet program will help to decrease their total body fat and decrease their incidence of prostate cancer.

The two most important measures in regard to the diagnosis of prostate cancer are: 1. A digital rectal examination, and 2. A Prostatic specific antigen blood test-PSA- done annually.

Prostate cancer in African-American men/Jamaican men is more aggressive and develops at an earlier age, making it much more difficult to treat. The take home lesson for men is to go to the doctor every year to have a digital rectal examination and a PSA blood test. Doing so will increase the chances of being cured of prostate cancer.

The total cost of prostate cancer in 2008 in the U.S. was 39.0 billion dollars.

Of all the different diseases that demonstrate the degree of the health disparity that exists in blacks compared to whites in the U.S., prostate cancer highlights the disparity more so than any other diseases. Clearly, the poorer socio-economic conditions of Black –American men and the constant racial stress they have to deal with plays a significant role in the incidence of prostate cancer.

Also important is the fact that blacks typically visit doctors for routine examinations less frequently than whites. When they do present to the physician for treatment, the cancer is often too well-advance to be treated adequately or successfully.

The lack of adequate health insurance or the absence of any health insurance at all also plays a significant role in the significantly poorer quality of medical treatment.

Prostate cancer in African-American men/Jamaican men is more aggressive and develops at an earlier age, making it much more difficult to treat. The take home lesson for men is to go to the doctor every year to have a digital rectal examination and a PSA blood test. Doing so will increase the chances of being cured of prostate cancer.

Another problem that men have to deal with is Benign Prostatic Hypertrophy-(BPH). The prostate gland is the size of a walnut and sits at the neck of the urinary bladder. The prostate gland has no health value. Its only value is that when a man ejaculates, the prostate gland discharges a liquid that facilitate the sperm to swim more easily to the egg for the purpose of impregnation. The sperm is quite capable of swimming toward the egg with help from prostatic secretion. All men will develop one problem or another from the prostate gland. Eighty percent of men 80 years of age or older suffer from BPH. The miseries that the prostate can cause for men include prostatitis, benign prostatic hypertrophy, prostate cancer, urinary tract obstruction, urinary tract infection, bleeding and possible kidney failure.

By age 30, the prostate gland begins to enlarge for reasons that are not altogether clear. It is possible that over stimulation by androgen plays a role in causing enlargement of the prostate gland. If stimulation by androgen is playing a part in causing the prostate gland to become enlarged, then diet has to be brought into the equation. The reason is that a diet rich in fats is known to increase the level of cholesterol in the blood. Androgen is a hormone and cholesterol is needed in the production of hormone in the human body. In fact, in the first chemical ring in the production of hormone, cholesterol is needed. That being the case, the higher the quantity of fat in the body, hence cholesterol, the more androgen the body contains. The more androgen the human body produces, the more stimulated the prostate gland becomes. The prostate gland cannot grow without stimulation from androgen. Thus, a high androgen level in a man's blood stream is associated with both increased prostate cancer and BPH.

It is a known fact that black men have a higher level of androgen than white men. This being the case, more black men has BPH than white men representing yet another example of disparity in a medical condition that afflicts blacks more than whites.

The Impact of Sexually Transmitted Diseases

Sexually transmitted diseases (STDs) are a major health problem in the U.S. As a group, blacks are affected more by STDs than Hispanics, Whites, Asians, and Native Americans. According to the CDC, about 19 million new cases of STD infections occur every year in the U.S. About half of these STDs occur among individuals 15 to 24 years of age. Among the reportable STDs are:

1. Gonorrhea
2. Chlamydia
3. Syphilis
4. HIV/AIDS
5. Human papilloma virus (HPV)
6. Hepatitis B
7. Hepatitis C

Chlamydia and Gonorrhea

"Chlamydia is the most frequently reported STD in the U.S. In 2007, 1,108,374 cases of Chlamydia were reported, up from 1,030,911 in 2006." The CDC estimates that there are 2.8 million new cases of Chlamydia in the U.S. every year; this means more than half of the new Chlamydia cases go unreported.

There remains an extremely high level of racial disparities among Blacks as compared to other ethnic groups in the percentage occurrence of STDs. For example, the January 2009 report put out by the CDC states that blacks, while representing "12 percent of the U.S. population, made up 70 percent of gonorrhea cases and almost half of all Chlamydia and syphilis cases in 2007. [48% and 46%, respectively]" Hispanics, according to this report, represent 15 percent of the U.S. population, and they account for 19 percent of all reported Chlamydia cases. According to the CDC, in 2007, black females 15-19 years of age had the highest Chlamydia rate of any group: 9,646.7 followed by black females 20-24 years of age: 8,671.5. The rate of reported Chlamydia per 100,000 was about eight times higher than that of white females. The rate in Hispanic females was three times higher than that of white females. The rates in American Indian/Alaska Native females and Pacific Islanders were lower than that of white females.[1]

---

[1] CDC, *Trends in Reportable Sexually Transmitted Diseases in the United States*, 2007.

Gonorrhea is the second most commonly reported STD in the U.S. There were 355,991 cases of Gonorrhea reported in the U.S. in 2007. Although the rate of gonorrhea remained stable in 2007, it is still found more frequently among ethnic minorities than among whites.

Both Chlamydia and gonorrhea have more serious health consequences in women than men. These two STDs cause pelvic inflammatory disease (PID) in women. PID causes vaginal discharge, lower abdominal pain, fever, tubo-ovarian abscess, and if left untreated, can cause infertility, spontaneous abortion, menstrual irregularity, breakthrough vaginal bleeding, excessive menstrual bleeding, painful menstruation, Fitzhugh-Curtis syndrome, etc. Fitzhugh-Curtis syndrome is a condition that develops because of chronic intra-pelvis infection resulting in adhesion. The adhesion causes violin like strings attaching to the outer liver capsule pooling on it, causing severe and chronic right-sided mid to upper abdominal. In addition, Gonorrhea can cause acute pharyngitis, proctitis, sepsis, septic arthritis, sub-acute bacterial endocarditis, etc.[1]

In men, three to four days post sexual intercourse with an infected sexual partner, burning and urinary frequency begins. Seven to ten days post sexual intercourse with an infected partner, a purulent discharge begins to come out of the penis with itching and pain in the groin. Sometimes, both Gonorrhea (GC) and Chlamydia cause no symptom and remain in a chronic indolent state. When this happens, the man is able to transfer these infections when he ejaculates into his sexual partner. Both men and women can carry these infections asymptomatically and chronically. In addition to urethritis, both acute and chronic prostatitis frequently develops as a result of GC and Chlamydia infections. Other complications of GC/Chlamydia in men who are infected with these micro-organisms include, sepsis, septic arthritis, SBE, pharyngitis, proctitis, urethral stricture, etc.[2]

---

[1]   To see how Chlamydia, Gonorrhea and PID are best evaluated and treated, See Alcena, Valiere, M.D., M.A.C.P., *The Best of Women's Health*, ISBN: 970-0-595-51058-0, iUniverse Publishing Co, November (2008).

[2]   To see how GC and Chlamydia are best evaluated and treated, see Alcena, Valiere, M.D., M.A.C.P., M*en's Health and Wellness for the New Millennium*, ISBN: 978-0-595-45782-3, iUniverse Publishing Co. (2007).

Genital Herpes

Genital herpes is the most common STD in the US. About 1 million individuals per year become infected with the herpes simplex virus. More than one in five men is infected with genital herpes simplex, for a total that adds up to more than 45 million Americans infected with the disease.

There are 2 types of genital herpes simplex, type 1, and type 2. The majority of genital herpes simplex infection is cause by herpes simplex type 2. Many people have herpes simplex infection and do not know it. Herpes simplex infection occurs across all racial and socio-economic groups.

The usual symptoms of genital herpes simplex in men are painful and itchy blisters at the head or near the head of the penis or rectum. The blisters eventually become painful ulcers or sores that usually about four weeks to heal. These outbreaks of herpes simplex type 2 can occur in the same areas every several months. Herpes simplex can be transmitted by anyone who is carrying it in his or her blood to his or her sexual partner during sexual intercourse.

Herpes simplex type 1 can also cause genital herpes and much more so than type 2. Herpes simplex type 1 usually causes fever blisters in the mouth and can be transmitted from mouth to genitalia during oral sex. Herpes simplex breakouts tend to occur during a time of stress. The reason for that is that during stress, a person secretes an excessive amount of adrenalin and elevated adrenalin causes a transient state of immuno-suppression which allows the herpes simplex virus to come out of its dormant state and causes breakouts to occur, usually in the spots from the previous break outs.

Syphilis Infection

Syphilis has historically been one of the most common STDs and has been in the new world since 1494 when Columbus and his men came to America. Syphilis is caused by the Spirochete Treponema Pallidum.

The rate of syphilis declined by 29 percent among blacks from 1997-1999. The rate remained stable among whites but increased by 20 percent among Hispanics. Overall, the rate of syphilis is 30 times higher in blacks than in whites; 125.4 cases per 100,000 blacks vs. 0.5 cases per 100,000 in whites. Syphilis is 50 percent more common in men than in women.1 "Over the past seven years, the rate of Syphilis in the U.S. has been increasing. In 2006 and 2007, the national primary and secondary rate increased 15.2 per cent from 3.3 to 3.8 cases per 100,000 people. The number of cases increased from 9,756 to 11,466."

The rate of primary and secondary syphilis among men is six times that of women. This is so mainly because of men who have sex with men.[2]

Syphilis is usually transmitted sexually. Once the organism is deposited into the human tissue it takes anywhere from fourteen to twenty one days for primary syphilis to develop.

The first manifestation of primary syphilis is a chancre which is often a painless sore. The chancre or syphilitic ulcer may be seen in the mouth, lips, arms, rectum, nipples, naval (belly button), etc. If left untreated, by four to eight weeks the chancre will heal spontaneously. Untreated syphilis spreads via the lymphatic system to disseminate throughout the body and cause secondary syphilis.

The first manifestation of secondary syphilis is usually a rash over the body. The rash can occur in the palms of the hands and sometimes under the feet. The rash is often scaly, smooth and may be itchy. This type of rash may resemble other rashes and may be difficult to distinguish just by looking at it. Because secondary syphilis is a systemic disease, it can cause sore throat, headache, fever, and weight loss with aches and pains. There may be large glands in the neck, under the arms and elsewhere in the body.

The physician has to be suspicious enough to order a blood test for syphilis. A scraping from the rash can be studied via dark field technique which might show the spirochetes treponoma pallidum wiggling around under the microscope. Similarly, "scraping material" taken from the chancre seen in primary syphilis when placed on a glass slide, covered with a cover slip, and placed under the microscope will show the spirochetes as

---

[1] CDC, *Tracking the Hidden Epidemic Trend in STD in the U.S.* (2000)

[2] CDC, *Trends in Reportable Sexually Transmitted Diseases in the United States* (2007).

well. Other diseases that secondary syphilis may cause include arthritis, hepatitis, condyloma, uveitis, iritis, otitis, CVA, kidney problems such as glomerulo nephritis, hepatitis, weight loss, poor appetite, memory loss, seizure, meningitis, etc.

After going through the primary and secondary stage in about twelve weeks, if syphilis remains untreated, it enters into the latent stage. Many people may have symptoms and signs of primary and secondary syphilis without recognizing them. In the span of about two years, if syphilis remains untreated, it enters into the tertiary stage.

Tertiary syphilis is an advanced stage of syphilis which can affect the heart, liver, brain, and other vital organs. The chronic involvement of vital organs with untreated syphilis is guaranteed to result in death. Syphilis can cause a person to become paralyzed and can cause a person to develop seizures and even insanity due to changes in the brain.

Two studies were done in the past that allow for an understanding of the effects of untreated syphilis on the human body. The first study was the Oslo study, which took place between 1891 and 1951. It included 2000 patients diagnosed clinically. The dark field test and Wasserman test were not yet in existence. Penicillin had not yet been discovered yet so there was no effective way to treat these patients. This study amply demonstrated the devastating effects of untreated syphilis on all parts of the human body, in particular, the brain, the aorta, the skeletal system, etc.

The second study is the infamous Tuskegee Study, which took place from 1932 - 1974 in Tuskegee, Alabama under the control of the United States Public Health Service.

From 1932 to 1974, under the leadership of the United States Public Health Service, 431 black men were injected with live syphilis organisms for the sole purpose of seeing what effects syphilis would have on their bodies. These men did not give consent for participating in the study. They did not know that they were being injected with live syphilis organisms.

The shameful reason given for this cruel, inhumane, barbaric, and racist study was to discover the effects of untreated syphilis on the human body. There were no scientific justifications for this study since the Oslo study had already shown what untreated syphilis could do to the human body.

In the history of the world, many events have taken place to catalogue men's cruelty towards each other: **slavery, the Holocaust, the Rwanda Massacre, and the Tuskegee Study** and now **Darfur** are a few such examples. In the Tuskegee Study, 431 men were sacrificed for no reason other than racism and bigotry, while the U.S. government, organized medicine, some black and white physicians, some white and black hospital administrators, and society stood by silently.

How could such a shameful, disgusting, and inhumane act against humanity have been allowed to take place? It was allowed to take place because those psychopathic bigots who set it up were insane.

Racism and bigotry are classic manifestations of psychosis. Those who are in charge of classifying psychiatric disorders have failed heretofore in their professional duties by failing to classify properly these racial behaviors for what there are. They are probably the most common forms of psychoses in the world. They were so in antiquity and they continue to be so today. All one has to do is look at atrocities that have taken place historically and the ones that are taking place today to see that these atrocities are the result of insane people who for one reason or another hate each other.

Just take a look at the killings, famines, rape of women that are taking place in the Congo and Darfur, and what took place in Bosnia, to name a few, and one will see that these are cruel, vicious and inhumane acts that groups of people are inflicting on others using race, religion and tribual differences as excuses.

Herpes Simplex

Herpes Simplex type 1 and type 2 are among the most common STDs known in medicine. In the U.S., about 45 million people twelve years or older are infected with genital herpes. One out five adolescents and adults are infected with this virus.

Genital herpes simplex type 2 is more common in women than in men. About one in four women and about one out of eight men have genital herpes. Type 2 genital herpes is more common than type 1 genital herpes. Type 1 genital herpes causes fever blisters mostly. It can be transferred to both males and females genitals during oral sex, kissing, or the virus can be transferred from hands to genitalia.

There is a great disparity in the rate of genital herpes in African-Americans as compared to whites. In one study, 48 percent of African-Americans were infected with herpes and 30 percent of Whites were infected with genital herpes.[1]

Human Papilloma Virus (HPV)

Human Papilloma Virus (HPV) is one of the most common STDs in the U.S. At any one moment, more than 20 million men and women are infected with HPV in the U.S. Every year, roughly 6.2 million new cases of HPV are diagnosed in the U.S. There are 100 types of HPV and 30 of them are sexually transmitted. HPV types 6 and 11 are associated with the causation of genital warts. The most common cause of cervical cancer is HPV infection.

The most common types of HPV that cause cervical cancer are types 16 and 18. HPV types 31, 33, 39, 45, 51, 52, 56, 58, 59, and 68 are also associated with cervical cancer. Worldwide, cervical cancer affects 470,000 women and 233,000 yearly die of cervical cancer. More blacks are infected with HPV than whites. In 2008, 11,070 women were diagnosed and 3,870 of these women died of cervical cancer. The incidence rates of cervical cancer in black women are about twice that of white women, and the death rate from cervical cancer is higher in black women as compared to white women.

This incredible disparity both in the rates of incidence of HPV and death of black women compared to white women has many causes including poverty, lack of health insurance, lack of health related literacy, cultural taboos, cultural habits, failure to visit physicians regularly, etc.

---

[1] Source: Centers for Disease Control

CHAPTER 14

DISPARITY IN HIV/AIDS IN THE U.S.

WHAT IS AIDS?

AIDS stands for Acquired Immune Deficiency Syndrome (as opposed to Inborn Immune Deficiency Syndrome). AIDS is referred to as Acquired Immune Deficiency Syndrome because the virus, the HIV Type I or Type 2, a retrovirus, enters the human body and attacks and kills the T helper lymphocyte (T4 or CD4), causing a decrease in their numbers, resulting in immunodeficiency of the body and in turn causing vulnerability to a multitude of diseases.

Some of these diseases are caused by the HIV viruses themselves and some of the diseases are caused by different opportunistic organisms that enter into the body at different times in the course of the HIV/AIDS syndrome. The T4 helper lymphocytes are in the body to help the body to be healthy, while the T8 suppressor lymphocytes are in the body to cause it to be sick when their numbers increase.

Therefore, in HIVAIDS, the number of T helper lymphocytes is lower than the number of the T suppressor lymphocytes, thereby inverting the T helper–to–T suppressor ratio.

How does the AIDS virus cause immunosuppression? Answer: The AIDS virus enters the bloodstream of the person being infected and quickly enters into the T cell CD4 lymphocytes. Once inside these lymphocytes, the virus multiplies by making copies of itself. Sometimes the virus can copy itself in numbers as large as a billion copies or several billion copies per day until the body gradually becomes more and more immunosuppressed, stage by stage, leading ultimately to full-blown AIDS and all of its associated problems and complications. All of these complications without treatment, or failed treatment, results in death.

AIDS - a historical perspective:

The first reported cases of AIDS appeared in an article published in June 1981 in *The New England Journal of Medicine*, in which a group of homosexual men was found to be sick with Pneumocystis carinii pneumonia. Further evaluation of the problem revealed that they were immunosuppressed and that their immunosuppressive state predisposed them to the development of Pneumocystis carinii pneumonia (PCP).

From that point on, the AIDS epidemic was underway. Subsequently it was published that a young man who was retarded and who lived in the streets of St. Louis, Missouri, who was a vagrant in the street of that city and who had frequent contacts with homosexual men, became very sick with an unknown disease associated with fever, weight loss, and pulmonary infection. He went on to die in the early 1960s from complications of the disease. After his death, an autopsy was performed and the pathologist wisely froze tissues and plasma that were taken from his body.

In the 1980s, after the AIDS epidemic was already underway, that pathologist evaluated the specimens that he had frozen, tested them for the AIDS virus, and found that these specimens were teeming with the AIDS virus. This documented that that young person in fact had died of AIDS. Therefore, in retrospect, the AIDS virus had been around in the United States, since the early 1960s, as documented by that early case. Many physicians in the US, including the author who, while in training in the inner city of New York City, saw many drug addicts presented to the hospital with febrile illness associated with large lymph nodes, etc. and had no idea what they had. When these lymph nodes were biopsied and examined by the pathologists, the reports were always lymphocytic hyperplasia - microscopically different materials used by heroin addicts.

These materials were said to be responsible for the so-called lymphocytic hyperplasia. Little did we know that most probably these men had AIDS; we simply did not know of the existence of the disease at that time. So, one does not have to go to Africa or to Haiti and other Third world countries to look for a scapegoat for the origin of the AIDS virus. The AIDS virus was in the inner cities of the United States, long before 1981 when the first cases of AIDS were published.

Be that as it may, blame passing, scape-goating, name calling, and finger pointing aside, AIDS is now worldwide and it knows no racial boundaries; it spares no social classes, it spares no sexes and it affects people of all ethnic backgrounds and religious beliefs. AIDS is the largest epidemic that mankind has ever known. Every few seconds a new person in the world is being infected with the AIDS virus and those infections are mainly being transmitted through sexual intercourse. As of the end of December 2001 there were 40 million cases of HIV/AIDS in the world and 28.1 million people in Sub-Saharan Africa live with the HIV virus.[1]

The Incidence of HIV/AIDS Worldwide and in the U.S:

Twenty-two million people worldwide have died of AIDS since the epidemic began in 1981. About 448,060 individuals have died of AIDS in the U.S. as of December 2000.[2] As of December 2001, the total number of AIDS cases was 40 million

There are presently an estimated 1,014,797 individuals living with HIV/AIDS in the U.S. Since the epidemic began in 1981, more than 565,927 individuals have died of AIDS in the U.S. In 2006, 56,300 people were diagnosed with HIV in the U.S. according to the CDC. World wide, in 2006, 39.5 million people were living with HIV (2.6 million more than in 2004). The number of new infections in 2006 rose to 5 million (400 000 more than in 2001).

Despite these startling figures in the U.S., the Sub-Sahara region of Africa remains the most affected region in the world. Two thirds of all people living with HIV live in this region—24.7 million people in 2006. Almost three quarters of all adults and children deaths due to AIDS occurred in sub-Saharan Africa—2.1 million of the global 2.9 million deaths.

---

[1] UNAIDS/WHO (Dec. 2001)

[2] *CDC, Morbidity and Mortality Report,* 50:432–434 (2001)

The number of people living with HIV increased in every region in the world in the past two years. The most striking increase occurred in East Asia, Eastern Europe and Central Asia, where the number of people living with HIV in 2006 was over one fifth (21%) higher than in 2004.

Access to treatment and care has greatly increased in recent years. Through the expanded provision of antiretroviral treatment, an estimated two million life years were gained since 2002 in low and middle income countries.

The centrality of high risk behavior (such as intravenous drug use, unprotected paid sex and unprotected sex between men and men and women) is especially evident in the HIV epidemics of Asia, Eastern Europe and Latin America.

Although the epidemics also extend into the general populations across the world, they remain highly concentrated around specific populations groups. According to the CDC" Racial disparity HIV/AIDS diagnoses remains high."

Blacks represent 13.8% of the U.S. population; yet, more than 50% of the newly diagnosed 184,991 cases of HIV/AIDS in 2001-2005 were blacks. Black men are seven times more likely to be diagnosed with HIV than white men are and black women are 21 times more likely than white women are to be diagnosed with HIV.[1]

In 2008, there were 33,000,000 people living with HIV/AIDS worldwide. In 2008, there were 22,000,000 people living with HIV/AIDS in Sub-Sahara Africa. In 2008, 2.1 million people died of AIDS worldwide.

Of the people who are infected with HIV/AIDS in the U.S., 73% are men and 26% are women. This includes children who are infected. Blacks make up 13.8% of the U.S. and yet, they make up 49% of those who are infected with HIV/AIDS in the U.S. Latinos make up 15% of the U.S. population and yet they make up 18% of those who have HIV/AIDS in the U.S. Whites make up 30% of those with HIV/AIDS in the U.S. 1% of Asian/Pacific Islanders have HIV/AIDS in the U.S. less than1% of American Indian/Alaska Native Have HIV/AIDS in the U.S. In the U.S, African American and Hispanics account for 67% of all HIV/AIDS cases and yet, they represent less 29% of the U.S. population.

Among women who are infected with HIV/AIDS, 64% are blacks, 15% are Hispanics, 19% are whites, 1% is Asian/Pacific, and less than1% is American Indian/Alaskan. According to the DC HIV/AIDS Administration, 3% of Washington DC has the HIV/AIDS infection.

In Washington, D.C., seven percent of African-American men are infected with HIV/AIDS and three per cent of African-American women are infected with AIDS. Fifty eight percent of these women got infected through heterosexual sexual intercourse with infected men and twenty five percent of these women became infected by using intravenous drugs. According to this recent report, Washington, D.C. is designated as "The Nation's HIV/AIDS Capitol". Source: DC HIV/AIDS Administration, March 15th 2009.

People receiving AIDS treatment worldwide:

According to the latest UNAIDS/WHO '3 by 5' data, more than 1.6 million people living with HIV were receiving antiretroviral therapy (ARV) therapy in low and middle income countries as of June 2006. This represents more than a four fold-increase since December of 2003. Overall, antiretroviral therapy coverage in low and middle income countries increased from 7% at the end of 2003 to 24% in June of 2006.

Money available in 2005 to provide for AIDS treatment:

In 2005, a total of U.S. $ 8.3 billion was estimated to be available for AIDS funding; this figure is estimated to rise to US$ 8.9 billion in 2006 and US$ 10 billion in 2007. However, it falls short of what was needed—U.S. $ 14.9 billion in 2006, US$ 18.1 billion in 2007 and US$22.1 billion in 2008.

For treatment and care, about 55% of these resources will be needed in Africa, 20% in Asia and the Pacific, 17% in Latin America and the Caribbean, 7% in Eastern Europe and 1% in North Africa and the Near East.

Why is AIDS so much more prevalent among people of color in the U.S. as compared to whites?

One reason more people of color are infected with the AIDS virus than whites is that more people of color are using intravenous drugs. Once infection with the AIDS virus occurs, the virus is quickly passed on to sexual partners. So, most of the HIV infections occurring in people of color result either from the use of IV drugs and/or from sex with their male sexual partners.

People may also get infected with the AIDS virus although they are not using IV drugs, but their drug-using sexual partners pass the virus onto them during unprotected sexual intercourse. Other reasons why the incidence of HIV/AIDS is so much higher in Blacks and Latinos are poverty, lack of health insurance coverage, multiple sexual partners, social denial, cultural taboos and lack of frequent visits to physicians to be tested for HIV.

AIDS - the clinical disease:

What are the different ways in which people can be infected with the AIDS virus?

1. Homosexual intercourse; men who have sex with men
2. Men who have sexual intercourse with both men and women
3. People who are injected with elicit IV drugs
4. People who have sexual intercourse with people who used elicit IV drugs
5. Women who have sexual intercourse with bisexual men
6. Individuals who receive blood or blood products contaminated with the HIV virus
7. Babies born to mothers who are infected with the HIV virus
8. Health workers who get stuck with needles contaminated with the HIV virus
9. Being bitten by an AIDS-infected person
10. Using the same toothbrush used by an AIDS-infected person
11. Engaging in passionate kissing with an AIDS-infected person
12. Engaging in oral sex with an AIDS-infected person
13. Sexually active uncircumcised men

What are some of the high-risk behaviors that can lead to the transmission of the AIDS virus from one person to another?

1.   Anal intercourse, men with women, or men with men.
2.   Intravenous drug use
3.   Prostitution, males or females
4.   Promiscuity, males or females
5.   Having unprotected sexual intercourse with strangers

In order for a person to become infected with the AIDS virus, the virus must enter the bloodstream of the person at risk.

How can a person become infected with the AIDS virus while having intravaginal intercourse with an infected woman?

The natural vaginal milieu of a woman has a high pH that allows for growth and multiplication of the HIV virus. Further, during sexual intercourse, there is also microtrauma of the capillaries that occurs as part of the natural events, making it possible for the HIV virus to enter into a person's bloodstream. The HIV virus is brought into the person's vaginal environment in the semen that is deposited within it during unprotected sexual intercourse. If a person were to have open sores, such as genital herpes, syphilitic sores, and other venereal chancres, etc., and have sexual intercourse with an HIV-infected woman or man, the person's chances of being infected increases several-fold, because the HIV virus can easily enter through these sores into the bloodstream. There is a very high correlation between STD and HIV infection.

Uncircumcised men have a higher risk of contracting HIV/ AIDS during sexual intercourse with infected partners because the foreskin of their penis is frequently *afflicted with balanitis, phemosis, paraphymosis, conditions* which allow for easy entry of the HIV virus into the blood stream. The different degrees of inflammation and breaks that result from these conditions in the foreskin of the penis create a perfect entry point for the HIV virus during sexual intercourse if one's sex partner is carrying the AIDS virus.

The following thesis is what this author proposed twenty-one years ago:

The notion that the AIDS virus had its genesis from Africa is a controversial topic. In my opinion, the data are not at all convincing as to where the virus originated.

It is my opinion that because the majority of men from Central Africa and Haiti are not circumcised, they constantly develop balanitis as a result of the heat and other problems, leading to breakage of the skin. This leads to chronic infections such as phimosis and paraphimosis. In this setting, there is frequent mini-ulceration of the foreskin of the penis. This represents an easy portal of entry for the virus during coitus with, let us say, an infected prostitute. Another possibility arises because the women in that part of the world do not shave the pubis. Thus there is the possibility of mini-lacerations occurring during coitus as the foreskin comes into contact with pubic hair. This is another possible portal of entry for the virus. This, to me, seems a more plausible explanation for female-to-male transmission in Central Africa and Haiti. [1]

(See reference below "AIDS in Third World Countries" below)

All other articles that talk about this idea as a possibility appeared after this author's article. Several articles have appeared recently, both in the scientific literature and in the lay press, reporting on research carried out in South Africa, Kenya, Uganda and elsewhere in Africa. This research has shown that when some men in these areas were circumcised, the incidence of HIV/AIDS transmission was decreased significantly: by 53% in Kenya and Uganda and 60% in South Africa. The World Health Organization (WHO) states that if followed world wide, male circumcision will decrease the transmission of HIV/AIDS by fifty percent. An article

---

[1] Alcena, Valiere, M.D., F.A.C.P., *AIDS in Third World Countries*, N.Y. State J Med 1986:86.446. See also *Infectious Disease News*, **Vol; 20, Number 3 (March, 2007)**; *Male circumcision for HIV prevention in young men in Kisumu, Kenya: a randomized controlled trial*, The Lancet 2007, 369:643-656, 24 February 2007.

in the January 14, 2007 of the New York Magazine calls male circumcision "A Real-World AIDS Vaccine".

This author takes full and deserving credit as the person who gave birth to the idea that male circumcision decreases the transmission of HIV/AIDS; this idea therefore is the author's intellectual property.

To date, this is the single biggest development since the IHV/AIDS epidemic was first recognized in 1981. There have been many major contributions but none is bigger than this one considering the fact that four million individuals yearly become infected with HIV/AIDS and 3.9 million individuals die yearly from AIDS worldwide. Male circumcision can cut both these numbers in half.

Notably, WHO, UNAIDS recommended male circumcision world-wide to decrease the incidence of HIV/AIDS transmission by 60% and to save 3 million lives. WHO, UNAIDS estimates that worldwide only 30% men are circumcised.[1]

The role of male circumcision in the prevention of HIV/AIDS has been extended to include Herpes virus type 2, Human papillomavirus, and syphilis.[2] As previously mentioned, 45 million people have herpes infection and 20 million people have HPV in the U.S. Syphilis is also on the rise, especially among men who have sex with men.

What happens when the HIV virus first enters into a person's bloodstream?

When the HIV virus enters the blood, the virus goes into the T helper lymphocytes, also known as CD4. Inside the CD4 lymphocytes that are in circulation, the HIV virus multiplies into millions at first then into billions of HIV virus copies per day. Within two to four weeks of the entry of the HIV virus into the bloodstream, the newly infected person often develops a flu-like syndrome with fever, general aches, chills, a runny nose, and even a cough, simulating acute rhinovirus or influenza infection. These symptoms quickly disappear and the person feels fine.

---

[1] Internal Medicine News, April 15, 2007.

[2] The New England Journal of Medicine, Vol. 360, No.13, pages 1298-1309, March 26, 2009.

The HIV viruses continue to multiply in the bloodstream and within the nodes of the person's body where they have entered. That represents the HIV stage 1 infection. During ten days to two weeks, the P24 antigen level becomes elevated. However, the HIV RNA PCR becomes elevated within about a week of someone becoming infected with the HIV virus, making it the earliest test and the most sensitive test that becomes positive, indicating the presence of HIV infection. The ELISA test becomes positive after the window period, which is from 6 to 12 weeks after infection. During the window period, the ELISA for the HIV, the P24 antigen and the HIV RNA PCR all will be positive, if the person is infected with the AIDS virus.

As the HIV viruses continue to multiply, the number of T4 lymphocytes decreases while the number of the T8 or T suppressor lymphocytes increases. That situation is what triggers the immunosuppressive states that occur in AIDS. As the infection progresses, the disease moves into different stages.

First, the HIV infection moves from the HIV-infected stage to ARC (AIDS-related complex) stage and then to the AIDS stage. The HIV stage may be completely silent, except for some patients who may develop thrombocytopenia (low platelet count) with or without enlarged nodes. The second stage is ARC. In that stage the person will start to lose weight with diffused lymph node enlargement, thrush in the mouth, diarrhea, fever, headache, oral hair leukoplakia, shingles, thrombocytopenia, molluscum contagiosum, recurrent herpes simplex, aphthous ulcer, condyloma, etc.

Some individuals take many years to progress from these stages to full-blown AIDS, ranging from eight to ten years, and still other individuals go quickly from these early stages to full-blown AIDS in four to six years. The mode of infection and the stage of HIV infection in the infector may play a role in how fast the infected person develops AIDS.

In this regard, there is a discussion in the literature regarding chemokine receptors CCR5 and CX4 and the role they seem to play in when certain individuals who are infected with the HIV virus progress to full-blown AIDS. Certain individuals who have some of these chemokine receptors in their blood may be resistant to the HIV infection. It would appear that there are different effects of the CCR2 and the CCR5 variants on HIV disease. One of the important factors is the overall makeup of the infected individual, in terms of his or her immune competence. Men with HIV/AIDS who have CCR5 chemokine receptors in their blood may develop resistance to some HIV/AIDS drugs. Recently, the FDA approved Maraviroc, which is a CCR5 co-receptor antagonist. Maraviroc taken at 150-300 mg per day can be used in adults with CCR5 HIV 1 circulating in their blood.

In order to say that a person has AIDS, clinically established criteria have to be met, as defined by the CDC. For example, a person with HIV infection whose CD4 count drops below 200 can be said to meet one of the criteria to have AIDS. The list of AIDS-defining illnesses includes the following:

CDC definition of AIDS:

To diagnose AIDS in a patient, the first blood test that is often done is the screening test called ELISA. If the ELISA test is positive, then the Western Blot Test is done to confirm whether the ELISA test is truly positive. The Western Blot Test is an actual electrophoresis of the protein contained within the body of the virus itself. The problem with the ELISA test is that it does not become positive until about eight to twelve weeks and it can be falsely positive. Another problem is that during that so-called window period, the HIV test could be falsely negative. To deal with that problem, the P24 antigen test can be done because it becomes positive within a minimum of ten days after the virus enters into the human body. In addition, the HIV DNA PCR test can be done to determine whether the HIV test is truly positive or not.

The take-home lesson for men and women who are involved in high-risk behaviors such as using IV drugs, having sexual intercourse with multiple sexual partners without the protection of a condom, and having sex with people whom they have just met without a condom, is to change these high-risk behaviors. For men who are not circumcised, they should consider seeing a urologist to get circumcised. These changes in behavior and medical treatments will decrease both the incidence and deaths from HIV/AIDS in both men and women.

CHAPTER 15

DISPARITY IN ARTHRITIS IN THE U.S.

WHAT IS ARTHRITIS?

Arthritis is an inflammatory condition that affects mainly joints resulting in swelling, pain, restriction of movement and, ultimately, deformity of the joints and bones. In addition, chronic bony destruction and edematous destruction also occur. However, certain types of arthritis at times can be multi-system, that is, affecting a multitude of organs such as the heart, lungs, kidneys, the blood system, etc.

In the United States, 46 million people in the U.S. suffer from arthritis, that is, 21 percent of the U.S. population has arthritis. That number is expected to rise even higher. By the year 2030, 40 percent of the U.S. population will suffer from one type of arthritis or another.[1] The annual cost of arthritis is 128 billion dollars.

The most common forms of arthritis in the U.S. are:

1. Osteoarthritis
2. Rheumatoid arthritis
3. Gouty arthritis
4. Ankylosing spondylitis
5. Psoriatic arthritis
6. Reiter's syndrome with arthritis
7. Systemic lupus erythematosus associated with arthritis
8. Polymyalgia rheumatica
9. Infectious arthritis

---

[1] Centers for Disease Control (CDC) and the Arthritis Foundation.

10. Lyme disease associated with arthritis
11. Sickle cell disease-associated arthritis
12. Fibromyalgia
13. Carpal tunnel Syndrome, etc.

Osteoarthritis

Osteoarthritis is the most common form of all the arthritides.

In the U.S., about 27 million individuals suffer with osteoarthritis arthritis. The mechanism that ultimately leads to osteoarthritis begins in the cartilage. The cartilage apparently releases a certain enzyme which, in time, causes breakdown of the joints to occur. Overtime, the areas in between the joints rub against each other, resulting in further bone destruction and deformities. Worldwide, about 60% of the population ages 60–70 have osteoarthritis of one joint or another. Certain ethnic groups seem to be affected with arthritis of some part of their body structure to a lesser degree than others. For example, Africans and southern Chinese have less arthritis in their hip joints. The knees seem to be the joint most frequently affected by osteoarthritis in all ethnic groups.

The aging process plays a major role in the development of osteoarthritis. In most cases, people over age 40 will develop osteoarthritis just by virtue of getting older. Some joints, such as the knees, hips, fingers, and spine are more affected by osteoarthritis than others such as the wrists and ankles.

Other commonly occurring types of Arthritis

In the United States, 2.1 million people suffer from rheumatoid arthritis, 3 million individuals suffer from gout, and 294,000 youngsters suffer from Juvenile Arthritis. At any one time, 59 million people complain of low back pain and similarly at any given time in the U.S., 30 million people are suffering from neck pain.

The following are illustrations of some examples of bony degenerative arthritic conditions:

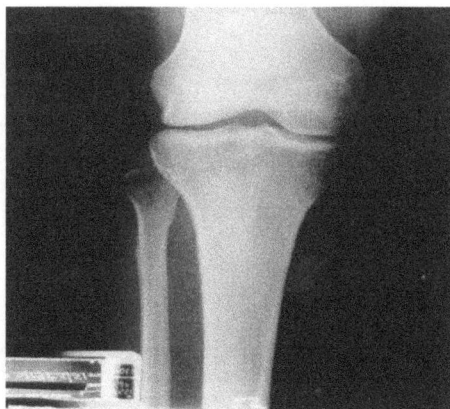

X-ray of a normal knee

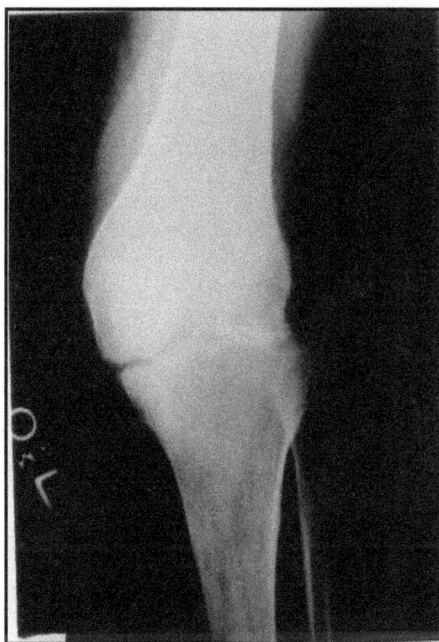

X-ray of a knee affected with osteoarthritis

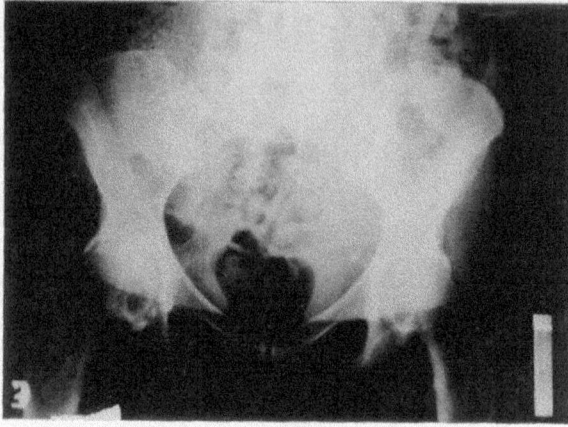

X-ray of hip joint affected with osteoarthritis

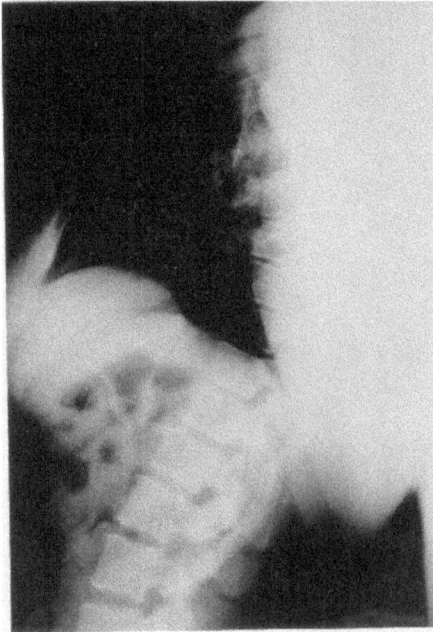

X-ray of lumbar spine affected with osteoarthritis

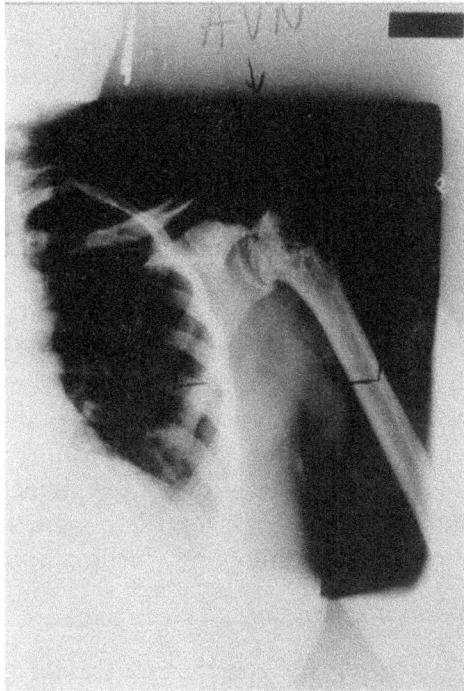

X-ray of shoulder joint showing aseptic necrosis in a patient with sickle cell anemia

Although the causes for osteoarthritis seem to be the same in all racial groups, there is a great disparity in the treatments that are provided to people of color as compared to whites.

More whites get joints replacement surgery than Blacks and Latinos and other minorities do. In particular, knee and hip replacements are offered several times more to whites than Blacks are. The main reason for this disparity is the fact that more blacks are uninsured than whites. Since more blacks are uninsured, they are less likely to seek medical care. Another reason for this disparity is that physicians are not anxious to offer operations to patients who are poor, have no money and have no insurance.

Systemic lupus erythematosus

Systemic lupus erythematosus (SLE) is an autoimmune disease of unknown cause that most commonly affects more women than men. There are roughly 1.5 million to 2 million people with SLE in the US.

According to the National Institute of Health, 9 out 10 individuals affected with lupus are women. SLE is more common in blacks and Hispanics than whites are. SLE is diagnosed three times more frequently in blacks than in whites. Here also, there exists great disparity in adherence to treatment in blacks as compared to whites. Blacks and Latinos even though there are affected more by SLE than other racial groups, are less likely to adhere to treatments. This failure to adhere to treatments, plus the natural course of SLE, results in some of the major complications that are frequently seen in Blacks and Latinos who have lupus. Among these complications are kidney failures and stroke.

Sarcoidosis:

The cause of sarcoidosis is not known. About 54,400 people in the U.S. suffer from sarcoidosis. This disease is more common in blacks than in whites. In the United States, 1 in 100,000 whites and 40 in 100,000 blacks have sarcoidosis. Sarcoidosis is twice more common in black women as compared to black men. However, there is a subgroup of young black men in the U.S.who have a very aggressive form of sarcoidosis. They also have a very high death rate. In general, blacks with sarcoidosis are diagnosed later than are whites because blacks often show up at the doctor's office with diseases that are more advanced than in whites.

Because, in the main, blacks are poorer than Whites are, they have less money to pay for health care services. There is no question that blacks are at a major disadvantage in every aspect of American society. However, the constant habit of not going to doctors on a timely basis to seek medical treatment has a strong cultural root as part of its genesis. The experience of slavery taught blacks to suffer quietly. To complain is to bring attention to oneself, which is a dangerous thing when you are living in bondage. To complain was to risk harsh treatments during slavery. In addition, many blacks still believe in old fashion remedies that are rooted in African cultural beliefs. Rooted in the black culture is the belief that to complain is a sign of weakness. Even now, many black men, in particular, feel that it is not "macho" to see a doctor when they do not feel well. Generally, black women are more likely to visit a doctor's office than are black men.

CHAPTER 16

DISPARITY IN EYE DISEASES IN THE U.S.

THE INCIDENCE OF EYE DISEASE and blindness is more common in minorities than in their white counterparts. Many of the diseases that predispose to the development of diseases in the eye are much more common in minorities. For example, diseases such as hypertension, diabetes mellitus, and glaucoma are much more common in minorities than in whites. The incidence of glaucoma is five times higher in blacks than in whites.

Glaucoma is the number-one cause of blindness among blacks. According to the American College of Ophthalmology, about 5 million Americans have glaucoma. Both hypertension and diabetes mellitus predispose a person to the development of glaucoma. Of the 5 million or so Americans who have glaucoma, a very large percentage of them are people of color. For reasons that are not yet clear, blacks have a higher propensity to develop glaucoma than whites do or any other ethnic groups. Glaucoma runs in families. That is to say, it is genetically transmitted. About 80,000 individuals go blind in the U.S. because of glaucoma yearly. What makes glaucoma so dangerous is the fact that it causes no pain. So, many individuals whose intraocular pressure are high, which is the first step to the development of glaucoma, do not know that the pressure inside the eye is high unless they go to the ophthalmologist to have their eye pressure tested.

As is the case for many other diseases, people of color often present for medical evaluations when the disease they are suffering from are already far advanced. Sometimes, these conditions have already reached too advanced stages to be helped even with the best medications or the best of medical procedures.

Pain does occur in the eye when the intraocular pressure rises acute to 40-50. This is an acute emergency requiring immediate treatment by an ophthalmologist to relieve the pressure inside the affected eye to prevent certain blindness.

There are four different types of glaucoma:

1. Primary open angle glaucoma
2. Secondary glaucoma
3. Angle closure glaucoma
4. Congenital glaucoma

One-fourth of all cases of glaucoma presents at birth and are due to congenital reasons. According to the Center for Health Statistics, in Bethesda, Maryland, 1.2 out of every 100 individuals have some form of eye disease. Although that is a high percentage, the incidence is much higher among men of color.

In adults, there are three different types of glaucoma:

1. Primary open angle glaucoma
2. Angle closure glaucoma, and
3. Low tension glaucoma

Glaucoma is seven to eight times higher in blacks than in whites. Blacks between the ages of 44 and 65 and, in particular, those who are hypertensive and have a family history of glaucoma have a 15 to 17 times greater possibility of developing glaucoma than do whites. According to published reports, glaucoma can be genetically transmitted and 30% of glaucoma patients have a family history of glaucoma.

Hypertensive retinopathy:

Hypertension has numerous complications associated with it and if left untreated will cause serious damage to occur in many organs. Prominent among these organs are the eyes. The increase in pressure within the vessels of the eye causes different degrees of damage to occur inside these vessels. The damaged vessels then trap platelets and other materials from the blood on the inner surface of these vessels, starting a nidus, which leads to plaque formation. It is also said that leakage of fatty material occurs out of these damaged vessels making the situation more complicated. That process perpetuates itself over time, causing different vascular abnormalities to occur inside the eyes, and resulting in hypertensive retinopathy.

Hypertensive retinopathy is graded as 1, 2, 3, and 4, depending on the severity of the vascular abnormalities.

Grade 1 shows arteriolar narrowing.

Grade 2 shows arterio-venous nicking, some exudates, and hemorrhages.

Grade 3 shows retinal edema, hemorrhage and cotton-wool spots.

Grade 4 shows a combination of Grade 3 plus papilledema.

The following images show examples of damage that can occur inside the eyes in people who suffer from hypertension.

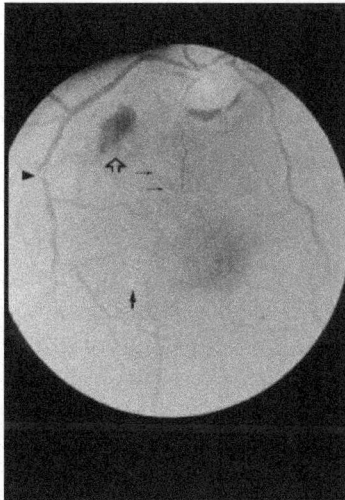

Figure 14.1: Showing different types of abnormalities in the eye of a hypertensive patient (hypertensive retinopathy). Small arrow showing silver wiring. Big arrow showing hard yellow exudates. Open arrow head showing blot hemorrhage. Arrow head showing A-V nicking.

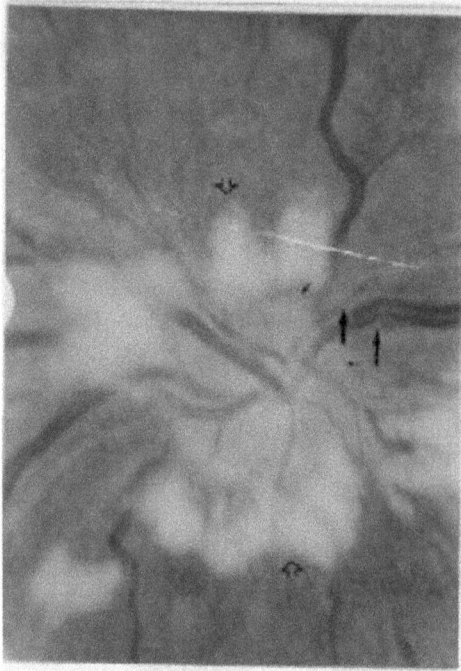

Showing different types of abnormalities in the eye of a hypertensive patient (hypertensive retinopathy). Small arrows showing early papilledema. One big arrow pointing to vein engorgement (larger vessel). The other big arrow pointing to arterial attenuation (smaller vessel); open arrow showing cotton wool exudates.

If proper treatment is not provided for these abnormalities, the patient often develops blindness. Hypertension is a very common disease, according to the latest estimates, occurring in about 73 million individuals in the United States. About 42% of these individuals, it is said, go untreated for hypertension. Hypertension is the number-one disease among blacks in the United States. The percentage of hypertension is higher among people of color than among whites.

Diabetes mellitus and its possible effects on the eyes:

Type II diabetes mellitus is very common among people of color and that is in part due to the fact that more than 60%

of Black Americans and 50% of Hispanics are obese. Obesity is highly associated with diabetes. According to the American Diabetic Association, there are roughly 24 million individuals diagnosed with diabetes mellitus in the United States and roughly 9 million more are undiagnosed. There are about 16 million pre-diabetics in the U.S. Each and every one of these individuals has the potential of developing diseases of the eyes including glaucoma and diabetic retinopathy with hemorrhage inside the eyes. Diabetes mellitus is 2-3 times more common in African Americans than among whites. In the United States, 13 percent of blacks have diabetes and 40-50 percent of blacks are at risk of developing diabetic retinopathy.

Diabetic retinopathy:

Diabetic retinopathy is a very serious disease which can cause blindness in a significant number of people who are diabetic. Some of the lesions that can be seen in patients who are suffering from diabetes mellitus include:

1. Micro-aneurysm
2. Arteriolar narrowing
3. Retinal edema
4. Hard exudates
5. Venous abnormalities
6. Soft exudates
7. Vitreous hemorrhages
8. Retinal hemorrhages
9. Retinal detachment, etc.

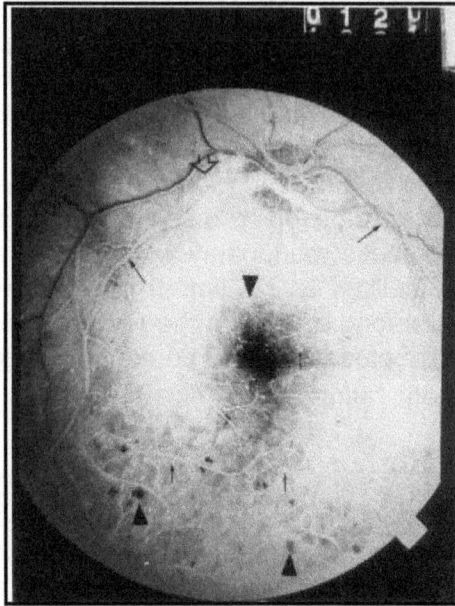

Showing different degrees of abnormalities in the eye of a patient with diabetes mellitus (diabetic retinopathy). Fluorescein angiogram shortly after injection of dye in patient's eye. Dye in arteries (white) and just starting to enter veins (large arrow). White area off NH is neovascular tuff (open arrow). White spots are hemorrhages (arrow heads). Tiny white spots are micro-aneurysms (small arrow).

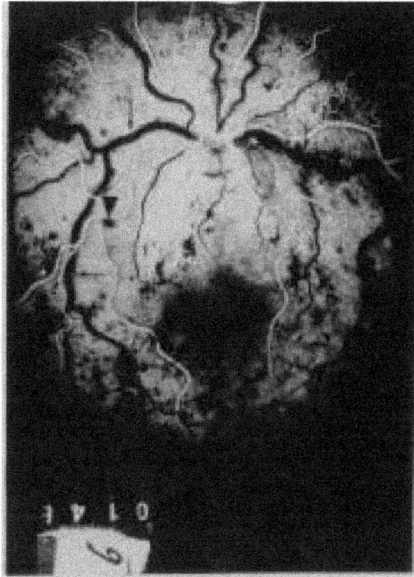

Showing different degrees of abnormalities in the eye of a patient with diabetes mellitus (diabetic retinopathy). Large arrows showing dilated veins. Arrow heads showing hemorrhages inside the eye.

It should be noted that eye symptoms and abnormalities may be the first signs that a person is suffering from diabetes mellitus. Very often, the patient presents to the ophthalmologist complaining of blurry vision and the examining ophthalmologist, if he or she suspects diabetes mellitus as a cause of the blurriness of the eyes, can then order the blood sugar to confirm whether it is elevated blood sugar that is, in fact, the cause.

As just stated, if the diabetic retinopathy is not very advanced, the fact that the blood sugar is elevated is enough to cause eye symptoms like blurry vision. Once the patient presents with symptoms of diabetes and is diagnosed with diabetes, the treating physician should refer the patient to an eye doctor for an appropriate eye evaluation to prevent unnecessary blindness due to diabetes mellitus. Because the incidence of diabetes mellitus is on the rise, diabetes-associated blindness is also on the rise. It is very important that individuals with diabetes mellitus understand that if they present themselves to the eye doctor early enough, keep their blood sugar under tight control, and remain under the constant care of a qualified ophthalmologist, they can prevent eventual blindness secondary to the effects of diabetes mellitus to the eyes.

Hemoglobinopathies and eye disease:

Sickle cell disease is the number-one abnormal hemoglobin disease that causes eye disease in those affected. Three different types of sickle cell diseases that can cause retinopathy are:

1. Sickle cell disease retinopathy (SS)
2. Sickle cell-C retinopathy (SC)
3. Sickle thalassemia retinopathy
4. Sickle cell trait (SA)

The most severe form of retinopathy among the three forms is that seen in sickle cell-C disease. There are two types of retinopathies seen in sickle disease: the proliferative type and the non-proliferative type. The proliferative type is more common in SC disease and sickle thalassemia than in SS disease.

The problems occur because of sludging of red blood cells inside the small vessels of the eyes. The red cells in sickle cell disease are mal-shaped and sticky, making it difficult for them to pass through these vessels. The result is occlusion of these vessels, resulting in a multitude of vascular abnormalities within the eyes.

The types of vascular abnormalities range from arteriovenous anastomosis and neo-vascularization that result in leakage of blood through these newly formed vessels and cause different degrees of hemorrhages. Retinal tear and detachment commonly occur as well. Fluorescein angiography is used to demonstrate these abnormalities. Photo-coagulation can be used as a treatment modality and laser is used to treat these conditions in the eyes of sicklers with retinopathy.

Clearly, Blacks, Hispanics and others who are afflicted with sickle cell disease have the highest propensity of developing blindness. In addition, many of these individuals also are frequently suffering with diseases such as hypertension, diabetes, and high cholesterol all of which can themselves cause blindness. Because blacks suffer disproportionately from all these diseases than do whites, the incidence of blindness is several times higher in blacks than it is in whites.

Sarcoidosis and its effects on the eyes:

Sarcoidosis, an inflammatory disease of unknown cause, is quite common in blacks and the eye is frequently affected because of that condition. In fact, eye symptoms are often the presenting symptoms of sarcoidosis including redness and swelling of the eyes with blurry vision. The ophthalmologist usually looks for anterior uveitis when examining a patient with suspected sarcoidosis. Slit-lamp examination is done to evaluate the eye when sarcoidosis is suspected. If sarcoidosis is not recognized and treated early with Prednisone, the result can mean total blindness in the affected eye. Glaucoma is also seen in chronic untreated sarcoidosis of the eye. The angiotensin-1-converting enzyme blood test is often elevated in individuals affected with sarcoidosis, and the serum calcium may be elevated as well.

AIDS and eye disease:

AIDS, as a viral illness, frequently affects the eyes. The most common infection that is seen in the eyes of AIDS patients is cytomegalovirus (CMV). CMV causes an infection of the eyes called retinitis. Blacks are affected with AIDS more commonly than whites in the United States are. As a result, AIDS-associated CMV retinitis is seen more often in blacks than in whites.

CMV retinitis in AIDS is quite difficult to treat and eradicate. The most effective medication is Ganciclovir.

In summary, eye diseases are more common in blacks than they are in whites. Diseases such as diabetes mellitus, hypertension, glaucoma, high cholesterol, sickle cell anemia, sarcoidosis and trauma in the workplace to the eyes, all of which contribute to the rising incidence of blindness seen in blacks as compared to whites in the U.S.

The overall economic and educational situations of the majority of blacks and other people of color in the U.S. will have to be improved drastically if it is expected that a real impact can be made to decrease their accelerated rate of blindness.

CHAPTER 17

DISPARITY IN DEPRESSION IN THE U.S.

Depression is one of the most common diseases that afflict us. It is said that in the United States, depression is as common as the common cold.

Every year, more than 19 million people in the U.S. suffer from depression. According to the National Institute of Mental Health, depression is the leading cause of non-fatal disability among individuals 15 to 44 years of age in developed countries like the U.S. and Canada. Another way to describe the incidence of depression in the U.S. is that 1 out of every 6 adults- 1 in 4 men and 1 in 10 women suffer from depression during their lifetimes. Worldwide depression is the second or third greatest cause of disability according to the Institute of Mental Health. The annual cost of depression in the U.S. is 53 billion dollars.

Primary care physicians write most prescriptions for anti-depression medications. Every year, 190 million prescriptions are written for anti-depression medications in the U.S. at a cost of 12 billion dollars.

There are different types of depression including:

1. Transient situational depression
2. Permanent or chronic situational depression
3. Depression associated with taking medications for a medical condition
4. Depression associated with alcohol or drug use
5. Minor classical depression
6. Major classical depression

7. Depression associated with anxiety reaction and panic attacks
8. Manic depression, etc.

Depression is three to four times more common in women than in men. In the United States, 12 million women suffer with depression yearly. One out eight women is expected to suffer from depression during their life time.

In the U.S., more than 17 million individuals (about 1 in 10 adults) suffer from a depressive episode at least once per year and more than 80% of the time these episodes go untreated.

Depression is more common in whites than in blacks. However, the disease is more severe in Blacks, Hispanics, and other minorities.

In any African ancestral society, mental illness is taboo because it is seen as a sign of weakness. A failure to be able to endure whatever it is that the majority of society can dish out and not only surviving it, enduring it and to be able to live long enough to tell one's children and grandchildren about it, is an essential part of the indigenous black culture. Blacks also see mental illness as a label that may be used to discriminate against them by the medical community, the job market, the legal community and by the law enforcement community to prevent them from getting ahead. Because blacks do not want any mention of mental illness on their records if they can help it, they hide their symptoms of depression and suffer in silence. When an interviewer tries to elicit symptoms of mental illness, they likely will not talk about it. They frequently do not tell even a physician/therapist about their mental illness unless they absolutely have to.

Blacks, as well as Hispanics, are extremely reluctant to go to the psychiatrist because of the fear of being labeled "crazy" in the case of blacks and "loco" in the case of Hispanics. Both these groups have culturally dealt with mental illness the way in which mental illness is dealt with in the African culture. In the African culture, when a person of African ancestry is troubled with a mood disorder, he or she goes to an elder or group of family members within the family for advice in order to deal with the problem.

Sometimes the group is organized as a committee, as is done frequently in developing countries. Never in this setting is the word mental illness used. In fact, in some third world countries, it is unlikely that a person will marry into a family that has an immediate member with a history of mental illness. As a result of racism and the different degrees and types of racial insensitivity and the suspicion and distrust that it causes, people of color in the U.S. are, for the most part, reluctant to seek help for mental illness, and in particular depression, because most of the psychiatrists-therapists are Caucasians.

The reasons people of color give for refusing to see Caucasian or non-minority psychiatrists-therapists include:

"Why should I go to the white and non-Hispanic, non-black psychiatrist-therapists to open up my mind, my soul, and my innermost secrets to them, when they belong to the group of people mostly responsible for my problems to begin with?"

It is a known fact that when people of color are evaluated for mental illness by white psychiatrists-therapists, the diagnosis is frequently wrong. The literature describes numerous cases of minorities who were diagnosed as schizophrenics when in fact they were not. Due to the subjectivity that is often involved in the diagnosis of mental illnesses, very frequently psychiatrist-therapists and others involved in treating people of color who have mental illness tend to give them a worse rather than a better diagnosis. Clearly racism, racial insensitivity, bigotry, and intellectual condescendence all play a role in this particular dynamic.

Transient situational depression occurs in all groups regardless of ethnic background. Normally, transient situational depression is a result of factors such as the loss of a job, a girlfriend, a death in the family, or the death of a close friend. In the case of some minority individuals, depression lasts longer because there is an underlying mental fragility born out of constant exposure to racial injustices which makes it easier to cross over the line to a more permanent depressive illness.

Anxiety and panic attacks are extremely common in people of color and occur more frequently than in whites since people of color typically have more reasons to be anxious and panicky. Minorities often face racism, poor education, lack of jobs, social isolation, lower economic status, harassment from law enforcement personnel and many other unspeakable injustices. All of these problems create a constant state of anger, disappointment, and uncertainty all frequently resulting in anxiety, panic attacks, and depression. Minority citizens in the U.S. may thus be described as the victims of the "STATUS SYNDROME".

What makes the whole situation even more serious is the fact that more often than not, minority men and women go without being diagnosed or treated for their depression for many years. In fact, the number of minorities who seek treatment for mental illness is quite low.

In the minority community, the primary care physician is likely to be the person most likely to see the vast majority of blacks and Hispanics suffering from depression. This is so because of the extreme resistance to the concept that they may, in fact, have a need for psychological care. This is a cultural fact and it must be understood, respected, and addressed with the greatest of care and sensitivity.

A major factor contributing to the lack of or poor mental health care that black, Hispanics and other minorities receive in the U.S. is the lack of health insurance. There are 46 million uninsured in the U.S. In 2007, 19.5 per cent black adults and 32.1 per cent of Hispanic adults were uninsured. With these very large percentages of the two largest minority groups uninsured, it is a given that minorities with mental health problems stand virtually no chance of finding mental health providers to care for them. In the United States, 2% of Psychiatrists, 2% of psychologists and 4% of social workers are black. The total numbers of Hispanic mental health professionals stands at 1,880 in the U.S. Blacks represent 13.8 % and Hispanics represent 15.1% the U.S. population. While there are about 900,000 physicians in the U.S., black physicians represent only 2% or 18,000 of the total number of physicians in 2008. With so few black physicians in the U.S., how can it be realistically expected that the black community could possibly receive the health care it needs?

In addition to the relentless torture of the mind of those afflicted with depression and other mental illnesses, many people resort to the extreme measure of committing suicide. The rate of suicide is many times higher in whites than in blacks. The suicide rate among white males is extremely high. The suicide rate among Hispanics is much higher than that of blacks. The suicide rate among white females is higher than that of black and Hispanic females. However, the suicide rate among blacks is on the rise. While black women attempt suicide four times more often than black males, black males complete their suicide attempts successfully three to four times more often than do black women.[1]

Culture plays a significant role in the significant differences that exist in the rate of suicide among the different racial groups. People of color have been taught from early on to expect to encounter difficulties in the journey of life. Therefore, they have developed all sorts of coping mechanisms to deal with hard times. Most whites on the other hand were never taught how to deal with hard times in life. Therefore, when they face with the difficulties in the journey of life, more often than not, many of them cannot cope and resort to drastic measures such as suicide.

It is imperative that psychiatric training programs include components of cultural diversity as it relates to minorities and the role of poverty and racism in particular. The incorporation of diversity training in all psychiatric training programs will help provide future mental health professionals the special skills and techniques necessary to address more successfully the problems faced by minorities and their diverse psychological problems. The directors of training programs in psychiatry, as well as psychology and social work and other related mental health fields can help to bring this about. The numbers of people of color in the mental health professions need to be increased significantly to help correct the disparities just described. In so doing, many of the barriers of mistrust may be lowered to a significant degree, making it easier for people of color to be more receptive to the care that the mental health professionals are able to provide.

---

[1]   Source: Centers for Disease Control

CHAPTER 18

DISPARITY IN ALCOHOLISM IN THE U.S.

OF ALL THE HABIT FORMING SUBSTANCES THAT are used in the United States, alcohol is number one. Alcohol use is widespread and alcoholism in the U.S. affects all segments of society. At any one time, there are roughly 18 to 21 million individuals in the United States receiving one type of treatment or another for alcoholism and its multitude of associated medical and psychosocial problems.

In the year 2004, 56.7% of men in the U.S. drank alcohol. Of this number, 47.1% men were regular drinkers of alcohol. The racial breakdown of men who drank alcohol on a regular basis in the year 2004 were white males: 63.3%, black males: 46,6 %, American-Indian and Alaska Native: 48.8%%, Asian men: 43.8%, Hispanic men: 61.7% mixed race men: 62%. In the year 2004 56.8% women drank alcohol in the U.S.: 61% White females drank alcohol: 38.6% American Indian females: 30.1% Asian females: 38.4% Hispanic females drank alcohol. It is estimated that in 1995, a total of 276 billion dollars were spent for the treatment of substance abuse in the United States, including alcohol.[1]

Alcohol use is one of the most serious medical problems. What makes alcohol abuse so easy and widespread is that a person does not need a prescription to buy alcohol. Alcohol is sold in liquor stores, bars, restaurants, airport shops, supermarkets, etc. Some men, and women, start drinking alcohol in their teens as a recreational habit or because of peer pressure. Frequently,

---

[1] CDC and National Center for Health Statistics, National Health Survey, *Family Core and Sample Adult Questionnaires,* 2004.

teenagers see their parents abusing alcohol at home, even getting drunk in front of them, and they come to believe it is alright for them to do the same thing (obviously it is not). It is not too difficult to see how some teenagers drink alcohol and are not reprimanded by their parents. The parents engage in the identical behavior in front of them. They then grow up thinking it is all right, and "cool" to do the same thing.

According to the literature, some forms of alcoholism is hereditary. Some alcoholic parents, it is said, transfer an alcoholism gene to their offspring, resulting in them becoming alcoholics as well. The evidence is quite compelling that it may indeed be so. On average, it would appear that whites start drinking at an earlier age than do blacks. Whites also on the aggregate have a higher incidence of alcohol abuse than do blacks do, but blacks suffer more from the physical effects of alcohol abuse than whites do.

"Blacks are 40% less likely than whites to develop alcohol abuse."[1]

For example, blacks seem to have a greater incidence of cirrhosis of the liver due to alcohol abuse than do whites. In fact, the death rate from cirrhosis of liver is two times as high for blacks as it is for whites. Generally, blacks suffer from more health problems because of alcohol abuse than do whites. Diseases such as cancer, hypertension, malnutrition, heart disease, diabetes, stroke, kidney failure, etc, are more prevalent in blacks who are alcoholics as compared to their white counterparts.

In 1990, it was estimated that 2–5 million older Americans had alcohol-related problems. The estimated hospital cost for hospital care for that group was in the neighborhood of 60 billion dollars.

Alcohol abuse is a multi-system disease. The following organs are most frequently affected by alcoholism:

1. The brain
2. The heart
3. The lungs
4. The liver
5. The pancreas
6. The breasts

[1] Med Page Today, April 22, 2208.

7. The gastrointestinal system
8. The blood system
9. The endocrine system
10. The mouth, throat, and esophagus
11. The neurological system
12. The psychological system
13. The skin
14. The genital system
15. Frequently, many of the organs and systems are affected in combination in the same person.

When a person drinks alcohol, the brain is the first organ to be affected. First, the alcohol is absorbed from the stomach into the bloodstream. Once in the bloodstream, the alcohol goes to the brain. The effect of alcohol on the brain depends on the level of alcohol in the blood and the length of time that the person has been drinking alcohol. The level of alcohol that causes drunkenness in one person is different in another person. In other words, different individuals respond differently to the effect of alcohol. The weight of a person determines how quickly he becomes intoxicated. Blood alcohol concentration (BAC) is measured as milligrams of alcohol per deciliter of blood. That same number can be converted in percent of alcohol concentration in the blood: 100 mg of alcohol per deciliter in the blood equals 100 mg percent or 0.1 percent of alcohol in the blood. For example, a person weighing 200 lbs. who drinks six drinks of hard liquor in one hour will likely develop a blood alcohol level of 100 mg per deciliter. A person who weighs 150 lbs. will reach a blood alcohol concentration of 100 mg per deciliter by drinking four drinks of hard liquor in one hour.

Alcohol is both a stimulant and a neuro-suppressor (brain suppressor). The first thing that happens when alcohol reaches the brain is a calming effect. It relaxes the person at first. Then, as more alcohol is consumed, a feeling of elation or euphoria ensues. Associated with that level of drinking is a mild form of excitement and the person may become talkative and giddy. This considered mild social drinking: 2–3 glasses of wine, 1–2 drinks of hard liquor, or 2–3 twelve-ounce bottles of beer. Further consumption of alcohol can create a state of drunkenness associated with excitation, rude behavior, and physical un-coordination. Another way of saying the same thing is that a standard drink of alcohol is usually expressed as a can of twelve ounces of beer, 1½ ounces of liquor—whiskey, vodka, etc., or five ounces of wine.

A usual dose of alcohol contains 13.7 grams of alcohol resulting in the following formulations of typical alcohol use:

5 ounces of wine =13.7 grams of alcohol
8 ounces of malt liquor =13.7 grams of alcohol
12 ounces of beer =13.7 grams of alcohol
1.5 ounces of gin = 13.7 grams of alcohol
1.5 ounces of rum =13.7 grams of alcohol
1.5 ounces of vodka =13.7 grams of alcohol
1.5 ounces of whiskey =13.7 grams of alcohol

Individuals metabolize alcohol differently and when taken with food in the stomach, alcohol absorption is slowed. If alcohol is consumed on an empty stomach, its full effects are felt more quickly.

Impairment due to the effects of alcohol occurs in a person when the alcohol concentration reaches 50 mg per deciliter. Women and elderly individuals show impairment from drinking alcohol at a lower concentration, probably 25–30 mg per deciliter. The risk of causing an automobile crash starts to occur when the blood alcohol concentration reaches 40 mg per deciliter. That risk rises when the blood alcohol concentration reaches 100 mg per deciliter. When the blood alcohol concentration reaches between 50– 70 mg per deciliter, most drivers are alcohol impaired and are unsuitable to drive. At that blood alcohol concentration, a person loses coordination (he or he cannot walk straight). At an alcohol level concentration of 100 mg per deciliter, a person has a more pronounced inability to walk, and would be stumbling around, and if that person attempts to drive, that person would be driving while drunk. When the blood alcohol concentration reaches 200 mg per deciliter, the person becomes confused, disoriented and may actually lose consciousness (alcohol blackout). When the blood alcohol concentration reaches 400 mg per deciliter, coma may ensue and death may occur.

What is alcoholism?

Alcoholism is a disease. A disease affects the human mind and the body. Alcoholism is a serious disease causing both psychological and medical complications of all sorts to the human body. Psychological dependence on alcohol is real and has devastating consequences on the affected individual and his or his family as well as society.

Like most addictions, it is very difficult to end an alcohol dependency. People who are addicted to alcohol crave it when they stop drinking. There are different patterns of alcohol dependency.

Some men and women drink alcohol in excess every day and feel the need to drink every day. A significant percentage of these individuals are able to go to work and function reasonably well on the job. They usually start drinking at lunchtime. They often consume two to three drinks with lunch, after work they will consume three to four more drinks at a bar on their way home or at job-related functions, and when they get home they will again have two to three more drinks with dinner. That is about ten drinks of hard liquor per day. On weekends, that number may quadruple. These individuals are what are called "functioning alcoholics". They work, earn money, and support their families. These individuals are found at all levels of society from the very rich to the very poor, and including the middle class. They are also referred to as people "who can handle their alcohol".

Working alcoholics must drink as soon as they get up in the morning to get going. They drink hard liquor at different times throughout the working day. Very often, people know that these people are alcoholics but tolerate them or cover up for them because they are oftentimes polite, very nice, and jovial, and when sober, they are productive at their work. Frequently, these people miss work because of heavy alcohol drinking and very often come up with very creative excuses as to why they were absent from work.

There are people who drink alcohol in large quantities and on such a regular basis that they become sick so frequently that they are unable to maintain a job. These are the hardcore, non-functioning alcoholics, who are entirely preoccupied with alcohol drinking on a daily basis. Such people can be found also in all segments of society, in all professions, in most religions and in all ethnic groups.

Alcohol use and peer pressure:

Peer pressure plays a significant role in alcohol abuse in teenagers due to peer pressure. Peer pressure also exists among adults to get together in a bar after work for a drink or two. Alcohol use often starts at an early age (from early adolescence to teenage years). Poor people drink alcohol for the same reasons that rich people drink alcohol—to socialize with their friends and to be less inhibited.

Eventually, dependency on alcohol develops, and once that happens, a preoccupation with alcohol drinking ensues resulting in alcoholism. It is not uncommon to find that many alcoholics grew up in broken homes where there was either a father or mother who abused alcohol, or that they have either a wife or girl friend who abuses alcohol. These individuals drink in order to please their significant others, or they drink because they saw their mothers or fathers doing it and they thought it was all right for them to start doing it also. Some of these persons are heads of household - single parents with children to bring up with no fathers around, and all the stress associated with running a household alone. In addition, all the problems associated with poverty, racism, and bigotry in a white-dominated society leads to heavy alcohol abuse in many minority groups.

The only differences that exist between the poor and the rich as it relates to alcohol abuse are that rich people have the financial means, which help them to be able to cover up their alcoholism. As for the upper-class, most are professionals, businessmen and women, athletes, entertainers and husbands, wives, sons, and daughters of the wealthy and privileged—all use alcohol for similar reasons. The exception is that they are not poor and undoubtedly suffer less discrimination (in fact, reasonably frequently they are the perpetrators of a significant degree of the discriminating themselves).

Because of their money, social and professional status, doors are frequently open for members of the upper classes that allow their alcoholism to appear more acceptable. The effects of alcohol on their bodies, however, is the same as that as the poorest people, because alcohol cares not how much money a person has, or for that matter, how privileged he or she is or how well a person is able to eat. The best caviar, the best cheeses, or the finest filets mignons in the world cannot protect the human body from the ultimate devastation of alcohol abuse.

Alcohol abuse affects the human body to the same degree, regardless of a person's nutritional state, with a few minor transient circumstances. For example, a person who is undernourished may go on an alcohol binge without eating. The fact that that person may have a low storage of carbohydrates in his or her liver from poor eating habits over an extended period, may cause that person to develop hypoglycemia (low blood sugar) more quickly than a person who drinks heavily, but has a better diet. That being said, the long-term toxic effects of alcohol abuse are the same in everyone who drinks alcohol heavily.

Although black people typically start drinking alcohol at a later age than their white counterparts do, the signs of alcoholism seem to appear earlier in blacks than in whites. Undoubtedly, poverty and racism play a major role in these differences. Poor people do not have as much money as the rich to go to the doctor for regular check-ups. Whites, generally, have more disposable income than minorities, and therefore whites, as a rule, get medical check-ups more often. Consequently, diseases such as cancer of the mouth and the throat, which are quite common in alcoholics, are picked up earlier in white alcoholics than they are picked up in poor minority alcoholics.

Alcohol is the most frequently used drug in the U.S. because it is legal to purchase it without a prescription. Alcohol is readily available and a person who wishes to indulge does not have to resort to illicit means to do so. States have established different age limits at which a young person can legally buy alcohol. On a yearly basis, alcohol use and its associated problems lead to more than 100,000 deaths in the United States. Of that number, close to 40,000 is due to cirrhosis of the liver and other associated medical complications of alcohol on the human body. The rest are due to alcohol-associated accidents on the highway (DWI) and homicides, etc.

The amount of alcohol that a person must drink on a long-term basis to cause damage to the liver is 80 grams of alcohol per day over an extended period ranging from 7 to 15 years. Eighty grams of alcohol can be found in a six-pack of 12 ounces of beer, because each 12-ounce can of beer has 13.7 grams of alcohol. If one multiplies that number, it adds up to 82.2 grams of alcohol. Some people drink twice that much beer per day. As stated above, if a person drinks this amount of alcohol on a regular basis, that person is guaranteed to develop alcoholic liver disease.

If a person drinks wine regularly, that is, a glass containing 3.5 fluid ounces of wine, consumes 9.6 grams of alcohol. A bottle of wine usually contains about five to six glasses of wine. People who drink two to three glasses of wine with dinner every night do not develop liver disease. It takes a minimum of 80 grams of alcohol per day on a regular basis over several years to develop fatty infiltration of the liver which leads to metamorphosis of fat, which then leads to necrosis, resulting in alcoholic liver disease with subsequent development of cirrhosis.

The same formula may be applied to drinking champagne. One glass of champagne contains 11 grams of alcohol; some champagne has 13 grams of alcohol per glass depending on how dry the champagne is. Therefore, it would take a tremendous amount of champagne consumption to add up to 80 grams of alcohol. About seven glasses of champagne daily over many years can cause a person to develop alcoholic cirrhosis. Fortunately, most people do not drink that much champagne.

However, if a person drinks martinis, that is a different situation. Each martini has 18.5 grams of alcohol. If he or she drinks five martinis per day, he or she is already drinking what is in fact in excess of the minimum amount of alcohol that is needed to cause liver disease. Five martinis equal 92.5 grams of alcohol. A Manhattan, for instance, has 19.9 grams of alcohol in it. Five Manhattans equal to 99.5 grams of alcohol. A gin Ricky has 21 grams of alcohol. Four gin Rickys equal 84 grams of alcohol. A High Ball contains 24 grams of alcohol. Four High Balls equal 96 grams of alcohol. A mint julep contains 29.2 grams of alcohol. Four mint juleps equal 116.8 grams of alcohol, etc. Therefore, it does not take many of these alcoholic drinks on a daily basis for a person to develop alcoholic liver disease.

The following is a brief description of the most frequently occurring types of damage to major organs affected by alcohol abuse.

Brain Damage

As described here, alcohol is very toxic to the brain. Acute alcohol ingestion alters a person's behavior by creating a state of excitation, restlessness, and poor social behavior. This agitated state leads to poor physical coordination, which frequently progresses to a state of drunkenness, leading to stupor and, at times, to coma. When intoxicated, alcoholics are a danger to themselves as well as a danger to others around them. The adult brain is affected by alcohol in many ways. For instance, people who abuse alcohol risk losing their ability to function properly on their jobs. They are likely to develop serious psychological problems which can cause disruption of their family lives, which often results in break-up of personal relationships such as marriages, etc.

Serious damage to the brain tissues leading to dementia is quite common in chronic alcoholics. Chronic alcohol abuse can lead to Korsakoff Syndrome, because of long-term vitamin B deficiencies. It is also known to be associated with acute episodes of encephalopathy, such as Wernicke's encephalopathy, caused by thiamine deficiencies.

Chronic alcohol abuse is also associated with other neurological abnormalities, such as ataxia (inability to walk in a straight line), altered mood functions with suicidal ideations, and peripheral neuropathy. Chronic alcoholics are prone to develop seizures, either because of alcohol withdrawal or because of recurrent traumas to the brain. Chronic alcoholics are often deficient in folic acid, all the B vitamins, magnesium, protein, and phosphate.

Liver Damage

Alcohol affects the liver because it is a direct toxin to the liver tissues. In other words, because alcohol is directly toxic to liver tissues, the amount of alcohol consumed, the frequency of that consumption, and the length of time an individual abuses alcohol, all determine the extent of the liver damage.

In some individuals, the damage to the liver occurs more quickly than in others, but one thing is certain, as long as a person abuses alcohol, his or her liver will be damaged by it. In some instances, the liver actually becomes swollen, which can in turn lead to acute enlargement of the spleen due to acute elevation of the portal pressure. The consequences can be acute rupture of the spleen endangering the life of the affected individual if it is not diagnosed properly and treated surgically quickly.

More chronically, however, alcohol causes tissues within the liver to become inflamed and the recurrent inflammatory reaction in time leads to scarring of the liver tissues, resulting in cirrhosis of the liver.

Once the liver becomes cirrhotic, a multitude of clinical problems can occur. The liver is needed to synthesize (produce) different proteins, which are necessary for good body functions. The liver is needed to make most of the coagulation factors required to prevent bleeding from occurring. The liver is necessary to store carbohydrates and to break down carbohydrates into usable sugars to use as fuel in the body. The liver is necessary to produce bile, which is needed to break down fats commonly consumed, plus a multitude of other essential functions. The liver is the largest organ in the body next to the skin and the skeletal system and it contains the largest supply of reticuloendothelial cells that are necessary to participate in the immune system, etc.

In addition, the liver is needed to help remove a multitude of breakdown products that the human body produces constantly. In so doing, the liver helps to detoxify the body. So, when the liver is sick and unable to produce needed materials for proper body functions, and is too sick to help remove waste materials from the body, the body becomes sicker. In other words, when the liver is too sick to function properly and fails, life cannot go on. Another way of putting it is that when the liver fails, the person dies. Alcohol abuse frequently causes the liver to fail.

Damage to the Spleen

Damage to the spleen is closely connected to damage to the liver. The spleen becomes sick in people who abuse alcohol chronically. Heavy alcohol intake causes cirrhosis of the liver. When alcohol damages the liver, that damage occludes the blood vessels that run through the liver. These damaged vessels cause narrowing and obstruction of the circulation inside the liver to occur.

Because of intra-liver obstruction of these vessels, over time, the pressure within the liver and the portal system rises. Portal hypertension then leads to enlargement of the spleen, resulting in a condition called hypersplenism. The enlarged spleen can at times become quite bulky, resulting in severe and chronic left-sided abdominal pain. The upper gastrointestinal system is quite frequently involved in that scenario and becomes quite sick because of the effects of cirrhosis of the liver and portal hypertension.

Because of the destruction and obstruction of these intra-hepatic and (intra-liver) circulation, the elevation of the portal pressure causes neo-vascularization (formation of new vessels) to occur (which is the body's way of trying to bypass the obstructed circulation in the intra-hepatic system). The new vessels, however, are superficial, meaning that they grow on the surface of the esophagus, resulting in esophageal varices.

Because these new vessels, called varices, are superficially located on the outer surface of the esophagus, they tend to rupture quite easily and bleed profusely. Esophageal bleeding is a major complication of cirrhosis of the liver with portal hypertension.

Pancreatis

The pancreas is another organ that is quite sensitive to the toxic effects of alcohol. Pancreatitis is a common complication of heavy and chronic alcohol use. It is not exactly clear as to the number of years a person has to abuse alcohol before his or her pancreas becomes sick. Some say that after seven years of alcohol abuse, pancreatitis can set in. Again, different people have different degrees of resistance and tolerance to the effects of alcohol.

Alcohol damages the pancreatic tissues, at first causing acute inflammation to occur. Inflammation causes marked swelling of the pancreas resulting in acute pancreatitis. After repeated attacks of acute pancreatitis, over several years, scarring of the pancreatic tissues occur, this in turn results in chronic pancreatitis.

Another scenario occurs when the chronicity of the pancreatic disease causes destruction of the pancreatic tissue, leaving empty spaces within the pancreas, causing pancreatic pseudo cysts to develop. Quite often, these pancreatic pseudo cysts become infected which in turn can lead to abscesses within the pancreas.

Still other sequelae of chronic pancreatitis is pancreatic failure, meaning the pancreas is so damaged that it is no longer able to produce the different enzymes that it was able to produce before it became damaged. These enzymes are necessary to aid in the digestive process of ingested fat. Failure of the pancreas to produce these necessary enzymes causes the development of greasy diarrhea to occur. To complicate matters further, when the pancreas fails, diabetes mellitus develops because the pancreas is no longer able to produce insulin for sugar metabolism. In addition, individuals who suffer from chronic pancreatitis experience constant and recurrent abdominal pain, nausea, vomiting, and diarrhea.

Gastritis

Another frequent complication of chronic and heavy alcohol use is gastritis resulting in upper gastrointestinal bleeding. Because alcohol is an irritating drug, it damages the superficial lining of the stomach, causing bleeding to occur, which at times can be quite severe and copious.

Damage to the Blood System

Another frequently affected system is the hematopoietic system (blood system). The effects of alcohol on the blood system are many and varied. Acutely, alcohol can suppress the bone marrow, resulting in the lowering of white blood cells, red blood cells, and platelets.

Chronic alcoholism can cause anemia because of recurrent upper gastrointestinal bleeding on the one hand, and on the other hand, alcohol abuse always leads to folic acid deficiency, resulting in folic acid deficiency anemia.

Many blacks already suffer from diseases such as sickle cell anemia which causes chronic hemolytic anemia, and when they abuse alcohol, that causes their anemic states to be much worse than in whites.

Ironically, alcohol abuse is more prevalent in the developed world where there is greater access to education and health information as well as more advanced health care. This is one area, where it appears that greater sophistication does not necessarily lead to healthier practices. One need not abstain from all alcohol use altogether and there may be benefits to be had from the moderate use of red wine, for example. Clearly, however, without alcohol abuse, one will have a healthier liver, heart, brain, pancreas, nervous, endocrinal, reproductive, blood and gastrointestinal systems, not to mention better emotional health and a healthier life overall.

CHAPTER 19

DISPARITY IN DRUG ADDICTION IN THE U.S.

ILLICIT DRUG USE is the second most common addictive habit in the United States. Alcohol is the number-one drug that is abused in the United States. According to the National Institute on Drug use, there are about 4 million drug addicts in the United States and about 3 million of them are addicted to cocaine. About 800,000 addicts are addicted to heroin. Marijuana is the most commonly abused illicit drug in the U.S. In 2006, 25 million individuals in the U.S. twelve years of age and older used Marijuana.[1] Several more millions used illicit drugs of different types.

In 2006, 560,000 people ages twelve and older, in the U.S. used heroin, 6 million people used cocaine, 1.5 million used crack cocaine, 73 million used tobacco products and many millions used Anabolic steroids.[2]

New estimates of the total costs of drug abuse in the U.S., which includes health and crime related costs, is in excess of half a trillion dollars every year. The annual cost of illicit drugs is 181 billion dollars, 168 billion dollars for tobacco and 185 billion for alcohol.[3]

Some of the most commonly abused prescription drugs include:

OxyContin
Demerol

---

[1] U.S. Drug Enforcement Administration (USDEA)
[2] National Institute of Drug abuse
[3] U.S. Department of Health & Human Services

Vicodin
Valium
Xanax
Nembutal
Darvon
Dilaudid
Librium
Ativan
Hydroconone
Ambien
Codeine
Morphine
Tranxene
Restoril
Methadone
Dexedrine
Ritalin, etc.

In 2006, 16.2 million people in the U.S. twelve years and older used prescription

Sedative, tranquilizer, stimulant, and pain killers for nonmedical purposes. Source: National Institute of Drug Abuse.

More people in the U.S. abuse prescription drugs every year than heroin and cocaine combine.

"Prescription drugs kill 300 per cent more Americans than illegal drugs."

Source: Natural News November 10, 2008

About 106,000 people abuse prescription drugs every year in the U.S.

The part of the brain stimulated by drugs that results in pleasurable feelings is the dopamine center, which is located at the base of the brain. Drugs such as heroin, cocaine, marijuana, opiates and amphetamines activate dopamine to release neurotransmitter substances, resulting in a pleasurable feeling called a "high", which drug addicts crave to experience. The dopamine center also functions to allow for the experience of sexual pleasure, enjoyment of foods, music, art, and beautiful things, and other aesthetic things that are pleasing to the ears and the eyes. Once an individual becomes addicted to any drug such as heroin, cocaine, crack-cocaine, marijuana, etc., he or she craves these drugs when the level of the drug decreases in the bloodstream. The craving for the drug oftentimes is quite painful.

Drug craving can lead to severe withdrawal symptoms such as sweating, headache, runny nose, abdominal cramps, diarrhea, poor appetite, insomnia, nightmares, etc. So, addiction to a drug, in particular cocaine, heroin and crack-cocaine, can drive addicts to do anything to get money to buy the drugs in order to satisfy the drug craving on the one hand, which is the more powerful and intense feeling, and to avoid going into drug withdrawal feelings.

On the other hand, once a person becomes addicted to drugs, it is very difficult to give it up. The addicted person becomes dependent on the drug and spends a great deal of time preoccupying himself or herself with finding money to get the next fix. He or she will spend rent money, food money, mortgage money. He or she will lie, steals, and commits crimes of different types and magnitude in order to get the money to pay for the drug. Frequently, he or she prostitutes himself or herself to get money to pay for the drug. Quite frequently, drug addicts commit crimes of different types to get money to support their drug habits.

Drug addiction is quite common among people of all racial and ethnic backgrounds and all social and economic Status. Wherever there is poverty and ghettos, there is always a high incidence of illicit drug used and trafficking. This occurs in part because of the state of despair that the vast majority people who live the ghetto feel. As a result, many of them turn to using illicit drug to ease their pain. Several generations of young Black and Latino men and women have been destroyed by the drug addition epidemics than began in the early 1960s and is still going strong unabated in the minority communities in the United States.

However, illicit drug used and trafficking have become prevalent in the suburbs of the United States as well, involving middle- and upper-class people. About 12.6 million people abuse prescription drugs for nonmedical reasons in the U.S. Statistically; more Whites use prescription drugs for nonmedical reasons than do Blacks or Hispanics and other minorities.

Percentage of drug use by race/ethnicity in the U.S.
American Indian 12.6%
Mixed race 11.8%
Blacks 9.5%
Whites 8.2%
Hispanics 6.6%
Asians 4.2%

Source: SAMHSA, 2007 Results of National Survey on Drug use and health Findings.

Of all the modern countries in the world, the United States has more people in jail at any one time than any other country. According to statistics, an excess of 2.3 million people are in jail in the United States. In fact, according to the Bureau of Justice Statistics, though Blacks represent about 12.2% of the United States population and Hispanics represent 14.3% in 2007. "Of the nearly 2.3 million people in jail in the United States, 2.1 million were men and 208,300 were women. Black males represented the largest percentage (35.4%) of inmates held in custody, followed by whites males (32.9%) and Hispanic males (17.9%)." Source: Bureau Justice Statistics, Prison Inmates at Midyear 2007 (Washington, DC US Department of Justice, June 2008).

In 2005, there were 253,300 state prison inmates serving time for drug offenses.

Among those serving jail time in 2005 for drug related offenses, 113,500 (44,8%) were black, 51,100 (20.2%) were Hispanic, and 72,300 (28.5%) were white". (Source: Bureau of Justice Statistics Prisoners in 2007)

(Washington, DC: US Department of Justice, December 2008.

Blacks are 6 times more likely to be imprisoned, and, Hispanics are more than 2 times more likely to be imprisoned than Whites are in the U.S. Two third of the people in jail in the U.S are Black and Hispanic males.

This disparity is yet another example of the many disparities that Blacks and other minorities face daily in the U.S. It is clear that there is one Justice System for Blacks and other minorities in the U.S. and a completely different Justice System for Whites In the U.S., it appears that the Justice System Considers Blacks and other minorities GUILTY until proven innocent. This same Justice System, appears to consider Whites INOCENT until proven guilty. In other words, if you are whites in the U.S. you right when you are right and you are right when you are wrong.

On the other hand, if you are minorities and you live in the U.S. you are wrong when you are right. If you happen to be minority and you have money, and you are accused of wrong doing, you are likely to be able to pay for legal services to prove your innocence; otherwise, you remain falsely guilty.

To the non legal observer, it appears that the Court System in the U.S. is set up to litigate legal Issues for Whites and to punish and jail minority men and women.

The economy of many white communities in the U.S. depends solely on the financial supports provided by jails that are builded in these communities. Therefore, these communities fight hard and lobby hard to keep these jails full with prisoners for their own financial gain and survival. People in these communities make their living at expense of the brutal miseries and the inhuman sufferings of these unfortunate people. No doubt, some of these prisoners are criminals who have committed crimes that have caused a lot of pain and sufferings to many people and their families and as such deserve to be in jail. The questions that remain to be answered are: Are there other ways to deal With those who have committed petty crimes rather than putting them in jail for extended period? Governments, ought to be able device some innovative ways to rehabilitate some of the petty criminals rather than jailing all of them. Legislation should be passed to allow judges the latitude necessary to sentence some of these people to lesser and less harsh sentences.

U.S. Society and Governments have the responsibilities to create the necessary environmental social, economic educational settings that would allow many those who are in jail an even chance to succeed in life. Doing so would have prevented the mental destruction of some of these individuals that ultimately contributed to their situations. No excuses are to be made for the horrible acts that these folks did that landed them in jail but, society in many ways created the breathing grounds that turn some of these people into criminals. Racial discrimination and the multitude of dehumanizing conditions it creates, set up the classic formula that leads to poverty, economic deprivation, poor education, high unemployment, poor housing, gang warfare, killings, drug trafficking, drug addiction and the chronic cycle of welfare dependency.

Some people become addicted to prescription drugs because of chronic pain associated with illness, such as cancer, arthritis, headaches, sickle cell disease, diabetic neuropathy and many other chronic diseases, which require chronic pain medication for relief.

Sometimes these individuals continue to get the prescription for these medications for a long time, but once they are no longer able to obtain these prescriptions, they resort to illicit drugs to ease their pain.

These are some of the many ways in which some of these people become chronic drug addicts. Drug addiction and all other addictions are psychological illnesses. The craving associated with drug addiction is controlled by neurotransmitters within the brain, in particular the dopamine center.

When the urge comes upon an addicted person to get a high, that person will do just about anything to get the money to buy the drug. Drug addiction is a mental illness and ought to be treated as such. Percentage-wise, IV drug addiction is more common among people of color, as compared to whites. This is so, because this type of drug addiction is more closely associated with the inner cities where most people of color live.

Although it is a known fact, the incidence of IV drug use is on the rise in the middle class as well as in the upper class communities of the United States. In other words, illicit drug use is also on Wall Street, Madison Avenue and most definitely in the suburbs.

In addition to the epidemic of drug addiction that is destroying the black community, the justice system treats blacks and other minorities differently than the whites.

It is utterly absurd and grossly unfair that an individual of color can be thrown in jail for life for using crack-cocaine , while individuals in the upper echelon of the United States society can be arrested for using cocaine or heroin, and get away at times with just a slap on the back of the hand. It would seem clear that these laws are placed in the books specifically to punish people of color, simply to get them out of society and throw them in jail. That seems to be grossly unfair, though no one should condone the use of drugs of any kind, because that is illegal.

However, putting it in its proper context, clearly there are inequities in the way the law is being used as it relates to people of color, versus people in the majority community. It is not hard to envision that these problems would be resolved rather promptly if the hardcore drug users in the United States were white middle class and the white community was being devastated by these drugs.

It is fair to say that a Marshall type of plan would have to be put into effect to deal with the drug problems, and as certain as the sun rises in the east and sets in the west, these problems would have been solved, if not completely, but certainly much better than they are being dealt with now.

Drug addiction in is a major problem in US society, and it contributes to the destruction that occurs in the families of those who fall victims to the awful power of drug addiction.

It is important for government to proactively undertake actions and create policies to get to the root causes of the drug addiction problems in all their aspects.

It is quite clear that present policies of building jails and throwing people in them and treating those people like animals are not working.

You cannot treat a psychological medical problem with jail or "three strikes, you are out" policies. Many politicians, district attorneys, and judges designed policies to deal with drug addicts that are popular with the electorate to get them elected at the expense of the people who are suffering from mental, medical, and physical problems of drug addiction. People who commit crimes ought to receive appropriate punishments for the crimes they committed. While in jail, treatments for their drug addiction and its associated problems ought to be provided to them. They should be given a real chance of rehabilitating themselves and they ought to be taught trades of different types that would enable them to be wage earners once they have completed their sentences and return to Society.

It is also important to realize that once a drug addict, always a drug addict and, that being the case, long-term psychological treatment ought to be made available to these individuals after they have left jail.

It is very costly to provide treatments for drug addicts. It is important that drug addiction treatments include prevention programs and educational programs. Programs to prevent the dissemination of drugs within the communities where these people live are very important. The federal and state governments need to spend the billions of dollars that are necessary to fund those programs to help them become successful.

It is important that drug addiction seminars begin at the earliest grades in schools across the country, so that children can be made aware of drug addiction and the ravages that it can cause to the human body and mind. It is important to let them know what the facts really are, so that when people approach them trying to get them involved in drugs, they can say "no". As it is right now, the incidence of drug addiction in schools across the country is high and begins at the earliest age, in elementary school up through middle school and high school. It is crucial that the educational system join forces with the government agencies to try to encourage drug prevention and drug education programs in elementary schools, middle schools, and high schools.

According to recent reports close to 25% of college students use drugs and alcohol. It is wrong, but they are at an age where they can make their own decisions. They also should be encouraged to give up drug use or not to start at all, because once a person starts using drugs, it is very difficult to give it up, because drug addiction is so overwhelming that these individuals are weakened by the force of the addiction.

The colleges also ought to organize drug prevention seminars on their campuses for the benefit of students and faculty. Confidential drug treatment programs ought to be offered and must be made easily available for college students who are using illicit drugs, to help them give up their habits. It is hypocritical to sweep it under the carpet and pretend that it does not exist.

To help individuals who are addicted to drugs, it requires money, it requires better governmental involvement, and it requires better involvement of the educators, the clergy, and other members of society working as a team to attack the scourge of drug addiction that is destroying so many people in the U.S. society.

CHAPTER 20

DISPARITY IN LUNG DISEASES IN THE U.S.

The most lung diseases that affect people in the U.S. are as follows:

1. Asthma
2. Chronic Obstructive Pulmonary Disease (COPD)/ Emphysema
3. Lung cancer
4. Sarcoidosis
5. Pulmonary embolism

Twenty two million people in the U.S. suffer from asthma and 7 million of them are children. Worldwide, 300 million people have asthma. Source: NIH Tuesday may 5, 2009.

In the U.S. the incidence of asthma is 4.4% in blacks and 4% in whites. Yet, the rate of asthma attacks and hospitalizations because of asthma are much higher blacks and other minorities. While some of the reasons for these differences are not known, some of them are well known. For example, many minority asthmatics live under conditions that cause the triggering of their asthma attacks.

Many minorities who suffer from asthma live in apartments where they are exposed to roaches which carry over their bodies irritants and other materials that get carried out by the air which when breathed by the asthmatics, precipitate asthmatic attacks.

Rats and mice which live in these apartments, no doubt, leave droppings that, when they get airborne, they also can be inhaled and participate in bringing about most asthmatic attacks.

In addition, minority asthmatics are exposed to the usual allergens in the same way as do whites, and, when these allergens are inhaled, they often precipitate asthmatic attacks.

Higher incidence of asthma are seen in industrialized areas where factories and industrialized machines emit sulfur dioxide and nitrogen oxide combining with air, which, when breathed, causes pulmonary diseases of different types, including asthma.

Upper airway infection is the most commonly associated precipitating factor triggering asthma attacks. The common infections that frequently precipitate asthmatic attacks are the common cold, which is brought on by rhinoviruses and parainfluenza viruses in adolescents, adults and in young children.

Other infections that can trigger acute asthmatic attacks are respiratory syncytial virus, bacterial infections, such as bacterial bronchitis, bacterial pneumonia, and bacterial sinusitis are all associated with the precipitation of asthmatic attacks. Different types of viral sinusitis, viral bronchitis, and viral pneumonias are also frequently associated with the precipitation of asthmatic attacks as well.

COPD is a combination of chronic bronchitis and emphysema or a combination of chronic bronchitis with asthma or emphysema with bronchiolitis. COPD is fourth leading cause of death in the U.S. after heart disease, stroke, and cancer.

There are more than 24 million individuals with COPD in the U.S. and roughly 12 million others with symptoms of COPD that are yet to be diagnosed. Worldwide, 293 million people have COPD/Emphysema.

Tobacco smoking is the number one cause of COPD/Emphysema in the U.S. and in the world.

The rates of smoking differ in different racial groups.

26.9% Native American smoke

11.2% Asians smoke

15.1% Hispanics smoke

21.8% Whites smoke

22.6% Blacks smoke.

Emphysema and Chronic Obstructive Pulmonary Disease in Men in The US: Adult men in the US who smoke by age and level of education are as follows:

Age 18 to 24	31.5 percent
25 to 44	30.3 percent
45 to 64	27.7 percent
65 and older	14.6 percent

38.4% of people with less than high education smoke cigarettes.

48.4% of people with GED diploma smoke cigarettes.

30.9% of people with high school diploma smoke cigarettes.

13.9% of people with college education smoke cigarettes.

9.1% of people with master degree, doctorate degrees and medical degree smoke cigarettes.

Source: National Center For health Statistics February 7, 2003, Center for Disease Control& Prevention.

According to the W H O, 1 person dies in the each 8 seconds as because of tobacco smoking.

In the U.S. each year, 440 deaths occur because of tobacco smoking. Smoking it is said reduces the smoker's live by an average of 12 years.

According to the American Heart Association, 47 million Americans smoke tobacco.

According to the W H O 15 billion cigarettes are sold worldwide every day.

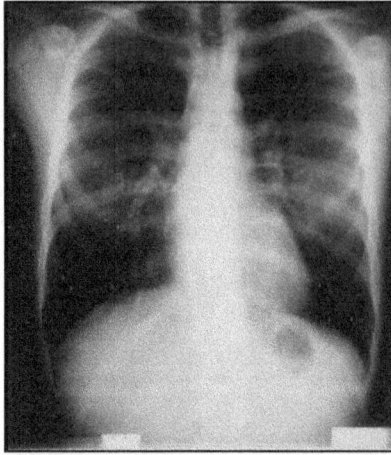

Figure 19.1–Normal chest X-ray in non-smoking female

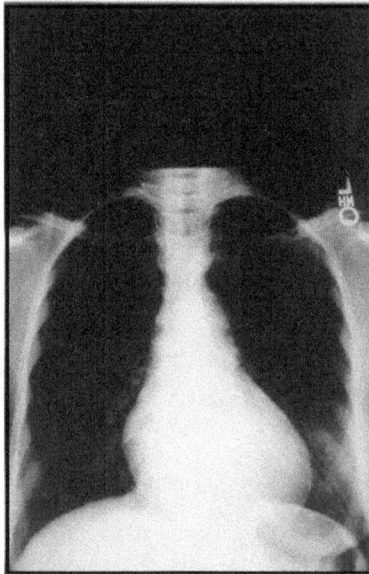

Figure 19.2– Abnormal chest X-ray in male smoker with emphysema

It is very costly to provide medical care those who suffer from COPD/Emphysema in the U.S. According to the National Heart, Lung and Blood Institute, in 2007, the annual costof COPD/ Emphysema was 42.6 billion dollars.

Tobacco smoking is the number one cause of lung cancer. In 2008, 215,020 individuals were diagnosed with lung cancer in the U.S. and 161,840 of them died of lung cancer.

In 2000-2003, more blacks developed lung cancers as compared whites in the U.S. During that time, 112.2 black men per 100.000 as compared to 81.7 white males per 100.000. had lung cancer. During that same time, 97.2 black males per 100.000 died of lung cancer as compared to 73.4 white males per 100.000 died of lung cancer.

In 2000-2003, 53.1 black women per 100,000 had lung cancer in the U.S. and 54.7 White women per 100,000 had lung cancer. During that same time, 39.8 black women Per 100,000 died of lung cancer and 42.2 white women per 100,000 died of lung cancer.

The main reason for the disparities both the incidence and deaths from lung cancer in blacks as compared to whites is because blacks smoke more tobacco than whites do and less black seek early medical care for symptoms related to lung cancer, thus, decreasing their chances of surviving the disease. Lung cancer occurs because smokers of tobacco inhale multitudes of carcinogenic chemicals that damage lung tissues causing cancer to develop in the lungs.

Some of the other common lung diseases that afflict people include:

Sarcoidosis:

Sarcoidosis is a very common inflammatory disease of unknown etiology that frequently affects the lungs 90% of the time. The most frequently affected ethnic groups are:
Blacks
Northern Europeans
Scandinavians
Irish Blacks in the U.S. have a 2% life time risk of becoming afflicted with sarcoidosis.

About 90% of people with sarcoidosis have their lungs involved with the disease process. Approximately, 50% of these individuals develop permanent lung abnormalities and about 15%

develop pulmonary fibrosis, and a significant percentage of these men go on to die of respiratory failure.

Pulmonary failure leading to death can be the result of pulmonary sarcoidosis in many people who suffer from sarcoidosis. It is clear, that blacks suffer disproportionately from sarcoidosis than any other racial groups, What it is also clear is that more blacks die of sarcoidosis than any other racial groups.

Tuberculosis is one of the world's most common infections that affect the lungs.

Worldwide more than 2 billion people are infected with tuberculosis. In 2007, according to the WHO 9.27 million new cases of tuberculosis were reported worldwide and 3 million people died from tuberculosis during that period. About 98% of these cases of tuberculosis occurred in the underdeveloped countries of the world.

In 2008, there were 12,898 cases of TB in the U.S. The majority of TB cases in the U.S. occur in foreign born people residing in the U.S. The incidence of TB in foreign born people is 10 times that of U.S. born people. The rates of TB among Hispanics and Blacks are eight that Whites. The rates of TB for Asians are 23 times higher than that of Whites in the U.S. The higher rates of TB seen in Blacks and Hispanics are in part associated with the high rates of HIV/AIDS that afflict these groups. Similarly, the high rates of HIV/AIDS that exist in the underdeveloped world account partly for the high rates of TB and deaths from TB.

No doubt, poverty, malnutrition, etc, that people many Blacks and other minorities experience on a daily basis, help to explain their higher propensity to the development TB.

About 10% of new tuberculosis cases that occur in the U.S. occur in individuals who se skin-test positive for TB are positive. Therefore that group of individuals serves as a reservoir for new TB cases.

The organism that causes tuberculosis is mycobacterium tuberculosis. Infected individuals can transmit the organism through droplets during coughing. Once the TB organism enters into the lungs of the person being infected, it locates itself in the upper/posterior part of the lung, where the oxygenation of that area of the lung favors its growth. The higher rates of TB seen among Blacks, Hispanics and other minorities both in the U.S.

and in world, clearly demonstrates the disparity that exists in the adverse effects that this deadly infection have on people of color as compared to whites.

The health crisis in the U.S. continues to have a negative impact on physicians, patients, hospitals, pharmacists and other health care related organizations. For example, a report just came out published by Health Affairs on May 14th 2009 and reported by American Medical News June 8, 2009 that shows that it costs physicians between 21 billion to 31 billion dollars per year for work they do for no pay to interact with HMOs and their many net works to get things done for patients.

In fact, the report shows that "The average physician spends 43 minutes per working day and more than three hours per week dealing with administrative requirements." Source: Health Affairs May 14th, 2009 (American Medical News June 8, 2009)

The HMOs create these administrative requirements, because it is extremely beneficial financially for them.

Still an other example of the financial greediness of the HMOs is the new report that shows While the yearly pay of some of the HMO's executives have going down, the yearly Salaries of the CEO of Health Net went up by 20% from 2007, $3.6 million in 2007 to $4.4 million in 2008. The salary of the CEO of Aetna went up by 5% from $23 million in 2007 to $24.3 million in 2208. The salary of CEO of Wellpoint went up by 8%. from $9 million in 2007 to $9.8 million in 2008.

All these things are going on while people are suffering from the poor economic situation. A lot of people are uninsured. (46 million people have no health insurance in the U.S.) a lot of people cant buy foods, a lot of people cant pay their mortgages a lot of people cant pay their rents, a lot of people cannot pay for their medications etc. The United States is the only country of the industrialized countries that does not have a single payer health insurance plan. Ideally, the single payer insurance plan would have been the preferred health insurance plan for the United States but this is not realistically going to happen.

Therefore, the time has come for the President and Congress to do what is necessary to create a Universal health

Insurance program and Public health plan to alleviate the sufferings that the present HMOs system is causing.

The public health plan that is being proposed is the only health plan that is designed to provide insurance coverage for the majority of the 46 million people with no insurance coverage in the U.S.

This plan will do the same things that the Medicare plan has so successfully done for so many million of Medicare eligible people. In addition, this plan will be less costly than all the other plans that are being talked about.

Medicare is run by the Federal government and it provides insurance coverage for more than 45 million elderly and disabled people. Medicaid is a health plan that provides health coverage for 60 million poor people. It is funded by the Federal government and the States.

Therefore, the Federal government already has many more years of experience running health insurance programs than do the HMOs and, clearly has run them better than do the HMOs.

FOR EXAMPLE, PRIVATE INSURANCE COMPANIES SPEND 14% ON ADMINISTRATIVE COSTS WHILE MEDICARE SPENDS ONLY 3% ON ADMINISTRATIVE COSTS.

The 11% difference in costs adds up to hundred of million of dollars of which goes in the pockets of highly paid administrators and their share holders of these private insurance companies.

Every one of the other health insurance plans under consideration, are nothing but a tricky way of reorganizing the present HMO/Drug company plans to continue to enrich themselves at the expense of patients, physicians, hospitals and other health care delivery entities.

The public is urged to support President Obama, health plan so that the 46 million uninsured can finally get health insurance which will enable them to receive the much needed and necessary health care to treat their medical conditions.

Moreover, the President's health plan will allow those who are presently insured to be better able to afford their health insurance through their employers. This health plan also proposes to help small businesses to be better able to provide insurance coverage for their employees, thus lifting the financial burden off their shoulders in a significant way. The President' health insurance plan will prevent the HMO's from picking and choosing who to insure. It will also allow people with pre-existing medical conditions to get health insurance. It will prevent the HMO's from refusing to pay medical bills for procedures and medications they feel are too expensive, even though their physicians feel that there are necessary.

The health insurance reform plan will save a lot of money because doctors won't have to practice defensive medicine by having too many expensive and unnecessary tests to avoid being sued by hungry and unscrupulous **LAWYERS** whose practice is to file frivolous law suits against doctors to make money for themselves and their clients.

The health insurance industry is spending million of dollars on TV advertising every day against health insurance reform, so that they can remain in business at the expense of the people of the United States.

It is grossly wrong and immoral what HMO's are doing. These folks and their supporters have no conscience. All they care about is the Status Quo so that they continue to enrich themselves while many people are dying because of lack health insurance coverage and many more are going bankrupt because of the high cost of health insurance.Health insurance reform is necessary for the health of the U.S. economy.

Presently the U.S. health care system costs 2.3 trillion dollars per year.

The money saved by cutting the waste that exists in the present system will help to improve the economic situation in the U.S. greatly.

President Obama's health insurance plan stands to benefit a lot of people. Congress ought to get to work quickly and effectively in a bypartisan way to pass this bill and send it to the White house for the president's signature.

About the author

Valiere Alcena is a practicing physician, medical scholar and, medical educator. He is clinical professor of medicine at the Albert Einstein College of Medicine Bronx N.Y. and Adjunct Professor of Medicine New York Medical College Valhalla N.Y. On May 15th 2008, Dr Alcena was inducted into the American College of Physicians as MASTER-MACP in ceremony held in Washington, D.C. Dr. Alcena is TV producer and TV Journalist.

He is the producer of the award winning weekly TV program Discussing Problems and Issues of Health with Dr Alcena (The longest running TV health show in the New Tri-State region, this show has been on the air since 1992) He is also the producer and host of the weekly TV program White Plains Community Health Fair Speaks

Dr Alcena is the Chairman of the White Plains Cable Access Commission. Dr Alcena Founded the Minority Students Affair Committee (MAC) at Albert Einstein College of Medicine in 1969 and is still the Chairman of that Committee. Dr Alcena created Community Health Fairs in the State of New York

He has published numerous articles in the scientific literature and on the Internet. Dr Alcena is credited as the physician who first came up with idea that male circumcision would decrease the incidence of HIV/AIDS. This idea has prevented several millions people from becoming infected with HIV/AIDS (6.3 million" between" 2006-2007). Close to 1 million deaths have also been prevented during that time because of male circumcision.

Time Magazine named the idea of "Male circumcision #1 among the Top 10 medical breakthroughs for the years 2007".

In 2007, Dr Alcena was named ICM teacher of the year at Albert Einstein College of Medicine.

The previous books that he has written include

1. The Status of Health of Blacks in the United States of America A Prescription for Improvement-1992

2. Third World Tropical Diet, Health Maintenance, and Medical Management Program 1992

3. The African American Health Book 1994

4. AIDS the Expending Epidemic, What the Public Needs to know A multi Cultural Overview 1994

5. African American Women's Health Book 2001

6. Women's Health and Wellness the Millennium 2002

7. Men Health and wellness for the New Millennium 2007

8. The Best of Women's Health 2008

Dr Alcena is recognized as the pioneer who first exposed the health disparity that existed and still exists in minorities in the United State of America. He wrote about it in 1992 in his first book.

Dr Alcena is the recipient of the PIONEER HEATH CARE DISPARITY AWARD
THE VOICE FOR THE ELIMINATION OF HEALTH CARE DISPARITY
PRESENTED BY
WESTCHESTER, PUBLIC, PRIVATE PARTNERSHIP FOR AGING SERVICES

ON JUNE 11th, 2005 at a ceremony held at Pace University Law School in White Plains N.Y.
In addition, Dr Alcena has received several dozen other awards over the years from many academic, governmental and, community organizations.

Dr Alcena has his medical office in White Plains New York where he practices General internal Medicine, Hematology and Medical Oncology.

For more about Dr Alcena, visit www.dralcena.com